Becoming

Becoming

Our Origins, Our Evolution, and Our Emergence as an Intelligent Species

Ralph D. Hermansen

ISBN: 1986153940
ISBN 13: 9781986153942

Table of Contents

Acknowledgments

Special thanks to my wife, Janet, who encourages and makes it possible for me to be a writer. Thanks also to Mark McGuire, whose computer skills have kept me going and solved countless problems. Dr. Tom Sutherland gave me guidance on the cosmology part of the book and James Morris, Tom Livermore, and John Lee gave me feedback on the manuscript.

Introduction

Overview

This book is about where we, humans, came from and how we got here. We begin at the beginning of time itself (i.e., the big bang), progress to the formation of the Earth, discuss the appearance of first life, witness the first multicelled life, progress through the geological periods, and track the evolution of man from ape ancestors. The later chapters are devoted to the evolution of primitive humans into the self-aware, intelligent humans now populating the world.

Human Ancestors Whom We Can Recognize

Fifty thousand years ago, our hunter-gatherer ancestors began to mentally reach beyond the needs of just surviving and began doing the things that we associate with "being human." We know about this because they left behind artifacts of that great cultural adventure. We are talking about people of our species, *Homo sapiens*, who began their migratory adventure in northern Africa and eventually occupied all the continents of the Earth except Antarctica. Yet it was in Europe that we see the clearest and most abundant evidence of this new enlightenment. There were new forms of hunting tools, nets and hooks for fishing, needles for sewing clothes, and semi-permanent dwellings, but the most impressive innovations were for nonessential items. These included bone flutes, carved statuettes, decorated spear throwers, shell jewelry, carved ivory beads by the thousands, ornate burials, and the most beautiful of all, polychromatic cave paintings of now-extinct animals.

This significant period is known as the "Great Leap Forward," and it reveals something not seen before, a people routinely using innovation to solve problems and a people using symbolism in every imaginable way. Despite the harsh Ice Age conditions, these people had moved far beyond mere survival living. They had free time for leisure and art expression. They also had time to seek answers to the big questions in life: "Where did I come from?" "How did we get here?" "Why must we die?" "Will the Gods answer my prayers?"

Our Special Place in Life

These kinds of questions are still with us today. We humans believe that we are special. After all, we are superior to all the other creatures on the planet in bending nature to our will. We live in houses that are heated in the winter and cooled in the summer; we can cover hundreds of miles in our cars in a single day or fly halfway around the world in an airplane; our mobile phones can connect us with others around the globe instantly; doctors fix medical problems today that would have killed us a few decades ago; and scientific progress is so rapid that even specialists are left behind in their own fields. So on one hand, we are becoming more powerful every day, but on the other hand, our special place in the universe has been shrinking at an equally colossal rate. In more primitive times, we visualized ourselves at the center of a modestly sized universe, made especially for us. However, that image has been changing fast as our understanding of the universe has expanded. The Hubble telescope has given us breathtaking images of distant galaxies. However, those distant galaxies were totally unknown to exist until the man Hubble discovered them in the early twentieth century. Until recently, it was pure speculation whether other stars had planets orbiting them. Now we know of hundreds of planets. Perhaps one of those planets has intelligent life on it too.

Cosmology

Evolution is a thread that runs through the fabric of this book. For many people, the term "evolution" evokes questions about the big bang theory and cosmology. For others, the term is strictly associated with the study of living things. In many minds, there is not a clear distinction between the two. Consequently, both topics are included in this book. In fact, the current scientific view of the origin of our universe is the subject of the book's first chapter. One might think of the instant of the big bang as time zero. We are talking about a time nearly 14 billion years ago. No stars or planets existed until millions of years after the big bang occurred. Our own star, the Sun, didn't even exist until 4.6 billion years ago.

If instead our Sun had been one of the first stars to appear in the early universe, that would not be good for us. For one thing, stars have a predictable life-span based on their size and temperature, and our Sun's life would have been ended by now. For another thing, there would have been no planets around our star. Elements heavier than hydrogen and helium would not yet have existed in that early universe, because such elements are manufactured in stars. When those stars explode at the ends of their lives, the heavy elements are scattered through space and incorporated in new stars.

Sometimes, people often tell me that they find the big bang theory difficult to believe. After all, how can something so huge, just pop into existence. I totally agree with them; the big bang theory seems to defy common sense. Yet science is driven by verifiable facts. We don't impose our beliefs on the universe. Instead we measure, observe, try to explain those observations, and make predictions from what we think we know. The more those predictions are verified, the greater our faith in our interpretation becomes. Part of chapter 1 will be a discussion of why we accept the big bang theory today despite its audacious concepts.

Evolution

Evolutionary Innovations

The remaining chapters of the book build on the concept of evolution (i.e., evolution of life). As we tell the story of life on Earth, it is really a story of life-forms building upon previous innovations. For example, amphibians were the dominant land animal at one time in the fossil record. However, they had to remain in the vicinity of water to reproduce. The innovation, which opened the continents for land animals, was the evolution of the amniotic egg. Here was an egg, which was not deposited in water and so the animals using it (i.e., Amniotes) could occupy the unclaimed lands. They became the reptiles. We and the other mammals are amniotes too.

Oxygen and Multicelled Life

Of course, that event was billions of years after the Earth first formed. The first half billion years or so on new planet Earth were an inhospitable time for life to flourish. The planet was simply too hot for life to survive. However, after it had cooled down enough, one-celled life-forms appeared, and they were the exclusive form of life until about one billion years ago when multicelled life-forms first appear. Why did it take so long for multicelled life to evolve? It turns out that an adequate oxygen presence was the determinant. There had to be sufficient oxygen in the oceans and air for multicellular life to exist. The newly formed Earth had virtually no oxygen. The oxygen we enjoy breathing today is generated by plants. Cyanobacteria were the original oxygen generator until plants evolved.

Prelude to Man

Chapters 2 through 4 take us from that newly formed, hot, inhospitable planet called Earth up to the time that bipedal apes appeared in Africa. We humans are more interested in our own story than we are in the creatures that came before us. Yet, I think it is important to at least present some of the vital aspects of this interval. One of those important lessons is that all living things today can trace their ancestry back to a universal common ancestor. This is a powerful concept and is borne out by the fact that all life-forms are based on DNA. It is also obvious to those who study nature. Charles Darwin knew nothing about DNA, and he wrote about the descent of all life from a common ancestor in the nineteenth century. Today, scientists compare the DNA of any two species and from that comparison, make good estimates of when that ancestor common to them existed. For example, through such comparisons we learned that we humans are most closely related to the African great apes (i.e., gorillas and chimpanzees). A more in-depth analysis showed that we were more closely related to the chimp than the gorilla and that we had a common ancestor five to seven million years ago. Finally, we are interested in the evolution of life from millions of years ago because we have thousands of genes that trace back that far or even farther. Life is parsimonious. It uses basic body-building plans over and over. We may be less than a hundred years old, but most of our genes are eternal by comparison.

From Ape to Man

Chapters 5 and 6 are devoted to following our evolutionary trail from common ape ancestor to where we had acquired a body much like ours today. Let us take a trip back in time: When we were that common ape ancestor five to seven million years ago, we lived in trees. Our ape bodies had evolved from monkeys, but we were now quite different. Monkeys run on tops of branches and have a tail to help maintain their balance. As apes, we swung under branches with our long arms and flexible collar joints. We lost our tails because we didn't need them anymore. Our ape brains are bigger and better than our monkey ancestors. For example, apes can recognize their reflection in a mirror whereas monkeys don't realize it is only a reflection. Times were really good for apes when the tropical forests were huge. Then times worsened as the trees began to disappear, and their source of fruits and nuts disappeared too.

The best response to their dilemma was to evolve into bipedal walkers, so they could spend more time on the ground searching for food. Chapter 5 tells us about these bipedal walkers and what their fossil record is able to tell us about them. If you have never heard of Lucy, whose 3.4-million-year-old skeleton survived the elements, you will learn about her and her kin. As time progressed, the heyday of the tree-dwelling ape did not return; in fact, things got worse. The bipedal apes had to look harder for food and shelter. Some turned to roots and fibrous vegetation and evolved huge teeth and chewing muscles. Another group, we call "Homo," took to scavenging meat and made stone tools to cut it. The meat eaters are our ancestors, and chapter 6 tells their story. These people underwent a remarkable transformation in a 2-million-year span. Their brain volume tripled, they lost their fur coats, their legs became longer, and their adaptation to living without tree shelter became successful. The most recent of them are us.

Development of Man

Chapters 7 through 9 follow the development of man from his long existence as a hunter-gatherer to the multifaceted social creature, which he or she has become. Once writing was invented, it was possible for the thoughts of someone living long before us to be read today. Writing gave us the magical ability to reach across time. However, writing is only about five thousand years old. How can we reach farther back than that? The field of archeology has filled in many of the gaps in the human-development story, but it depends on finding bones, tools, pottery, and other relics of the past. The study of languages is another tool that lets us look into the past. In the same manner that new species arise or disappear, so new languages rise and disappear. The farther back we attempt to trace this lineage of languages, the harder it is to do. What would be ideal is a new tool, which can reach deeply into our past and help connect all the factoids from the other research tools. It turns out that there is such a tool, and it became usable only in recent decades. Chapter 7 tells that story.

In the nucleus of our cells lies a molecule containing a four-letter code, and the secret to life. We are talking about our DNA. It contains a lot of information about our past because it is a huge molecule. To illustrate how big, think of it this way: the molecule could be visualized as a long pearl necklace—3.2 billion pearls in length—a long necklace indeed! Biochemists call those pearls, nucleotides, and they are in reality, much tinier than pearls and composed of organic matter, not mineral matter. My point is that we have an enormous molecule at our disposal, one that has recorded our entire history going back to when we lived in trees. In chapter 7, we discuss how science has unlocked the secrets stored within that molecule and what we have learned from it. Our species, *Homo sapiens*, has occupied the entire Earth for thousands of years. Where did we originate, and how did that journey from the original location take place? We think we know the answer to that question in surprising detail, thanks to DNA. Were there other human species around then, and did we ever mate with them? It turns out we did. We even discovered a species previously unknown to us called Denisovan man.

Chapter 8 examines the progress of human intelligence and what factors influenced its course. The multimillion-year transition from vegetarian tree-dwelling ape to meat-eating hunter-gatherer was a major part of it. The fabrication of cutting instruments from stone cobbles required an advancement in mental imaging, manual dexterity, and self-discipline. The acquisition of verbal communication accelerated the need for brain reorganization. Finally, the power of sexual selection, the advantages of pair bonding, and changing cultural influences shaped the modern human into his current form.

Chapter 9 examines a dynamic period for humans, that of the agricultural explosion and the rapid advance into villages, cities, and finally empires. The Fertile Crescent has been deemed to be the center from which civilization first emerged, yet it also independently emerged in Mesoamerica and Africa. Agriculture fostered larger populations and division of labor. An innovative culture already in place from the previous Great Leap Forward, only operated at a faster pace.

Chapter 10 looks to where humanity is headed. The question as to whether humans continue to evolve is controversial. However, there is no doubt that favorable mutations are being generated at a record pace. The Red Queen Effect guarantees that human's battle against evolving microbes and parasites will continue forever. Meanwhile science and medical technology forge ahead taking us closer to the half-man and half-machine cyborgs of science fiction and taking us closer to reengineered humans via genetic modification.

One

The Big Bang, Stars, and Our Solar System

The Chapter in a Nutshell

Only we humans think about our origins, and that picture has undergone considerable revision due to scientific progress. Early man thought the Earth was flat and that the Sun and other heavenly bodies were gods looking down at us. Thinking became more sophisticated with time and yet, new concepts like a spherical Earth, a solar system, or a Milky Way galaxy were accepted only grudgingly. In the seventeenth century, Isaac Newton taught the world that there is predictable order to the universe. Accordingly, his followers favored the Steady-State Model, which described a universe that always had existed and always will exist. Alas, this image was disrupted three centuries later. It was Albert Einstein who upset the Newtonian order with his theories of special relativity and general relativity. Concurrently, the astronomer Edwin Hubble caused us to change our belief that our galaxy was the extent of the universe, when he discovered that galaxies existed beyond the Milky Way. Other astronomers began looking for them and learned that there are actually billions of them. Moreover, these galaxies were seemingly moving away from each other, or was it that space itself was expanding and carrying these galaxies with it? Georges LeMaitre measured these changes and imagined the expansion in reverse to a time when all matter condensed to a point. This was the genesis of the big bang theory, which is accepted by most astrophysicists today. Although it is hard to imagine that an entire universe could come from nothing, two facts make the big bang theory hard to dispute. First of all, theory predicts that only three elements are formed from the hot plasma, namely mostly hydrogen, some helium, and a trace of lithium. This is exactly what is observed in the universe. All other elements are manufactured in stars. Second, the big bang is calculated to have happened 13.7 billion years ago. The cosmic background radiation agrees with that estimate.

Stars formed within the first billion years after the big bang. Now, stars go through a life cycle of being born, aging, and dying. The aging occurs as the hydrogen that fuels them gets consumed. Consequently, many star life cycles had occurred before our Sun was born. In fact, our solar system was born only 4.6 billion years ago, long after the first stars came into being. Eight planets orbit our sun. It used to be nine until Pluto was demoted to dwarf planet, but now we learn that an unknown ninth planet might exist in an extreme orbit. Our own planet Earth is strongly influenced by its large moon. This earth-moon system is thought to be the result of a major impact long ago.

Man's Expanding View of the Universe

How far back can we go in time? Was there a time when everything began? Our big bang concept of the universe is radically different than it was a few centuries ago when we pictured a universe that always existed and would always exist. And think how different the Steady-State Model is from that of our ancient ancestors, where the Earth was the only world and the Sun, Moon, and stars were lighted objects of wonder and awe. As early as the fifteenth century, most people thought the Earth was flat. Christopher Columbus was considered bold for sailing west in 1492. Some feared he might sail off the edge of that flat world. Today, we know that the Earth is one of the tiny rocky planets orbiting an average star. There are some giant planets orbiting that star too. However, they orbit much farther out from the star than we do. Like our planet with its large romantic moon, many of the other planets have moons orbiting about them too. Altogether, the central star, which we call the Sun, the planets, and the moons comprise our solar system. In 1610 Galileo Galilei trained his primitive telescope on the planet Jupiter and observed four bright points of light all lying on a straight line, which bisected the planet. Night after night he recorded their positions and deduced from their motions that they were actually moons orbiting the planet Jupiter. Alas, if Galileo was expecting praise and recognition for his scientific discovery, he was deeply disappointed. Instead, he was jailed for heresy and threatened to have his properties stripped from him if he did not recant.

Galileo is honored for his contributions in modern times though. In his time (1564–1642), he did pioneering work in physics, mathematics, and astronomy. He was an inventor too. Galileo was the first to observe the Milky Way galaxy stars with his telescope, but it was Immanuel Kant, who in 1755, proposed that all these innumerable stars were collectively held together by gravity. So, we see that the concept of us being part of a huge star-filled universe is fairly new.

A Galaxy-Sized Universe

Today, we know that our Milky Way galaxy is immense and contains about two hundred billion stars. The diameter of the galaxy is one hundred thousand light years. In order to appreciate this immense size, consider that a light year is the distance that light travels in a year, and that the speed of light is 186,000 miles per second. So multiply 100,000 years by 36,500,000 seconds per year by 186,000 miles per second, and you have the number of miles to cross our galaxy. Albert Einstein showed that no mass-containing thing can go as fast as the speed of light, but presuming that we could travel that fast, it would still take us one hundred thousand years to traverse our galaxy. In other words, our galaxy is so huge that it strains our imaginations. It is not hard to believe that this immense galaxy is the entire universe, and people did believe that until a man named Hubble came along.

A Universe with Countless Galaxies

Now what if I told you that there are one hundred billion galaxies in the universe? It is hard to conceive of something that huge, isn't it? In the 1920s, an astronomer named Edwin Hubble was using the one-hundred-inch Hooker Telescope at Mount Wilson Observatory, when he realized that he was observing stars outside of the Milky Way Galaxy. This discovery was breathtaking at the time because until this discovery, everyone believed that the Milky Way Galaxy was the entire universe. Suddenly, our concept of the universe changed radically. It was far, far bigger than we had ever imagined. After Hubble's discovery, astronomers turned their attention toward finding new galaxies and describing them. For example, our neighboring galaxy, the Andromeda galaxy, was found to contain even more stars than our own galaxy. Today, enthusiasts at telescope outings like to point out the Andromeda galaxy on extra-clear nights. I have been atop Mount Pinos in California, with them, and I have seen it without the aid of a telescope. It occupies about the same-sized space in the sky as the Moon but it is very faint. Today, the Hubble telescope makes it possible to observe countless galaxies.

Modeling the Universe

The Steady-State Model

Isaac Newton (1642–1726) was a natural philosopher, which in his time meant that he was multidisciplined. He was a mathematician, astronomer, physicist, and general scientist of extraordinary brilliance. He changed an unpredictable world into one that had stability, harmony, and order. His laws of motion are a case in point. His classical mechanics is the foundation of any physics course. In Newton's universe, the length, width, and depth of space were invariant entities, and time ticked forward in a predictable manner too. His viewpoint had widespread influence on the thinkers of the world. The universe was considered to be orderly too. The Steady-State Model assumed that the universe was unchanging. It had always existed and would always continue to exist, ticking away like a very dependable celestial clock. That was until a man named Einstein came on the scene.

Einstein and a New Vision

The field of science that is concerned with theories about the origin of the universe is called cosmology. Cosmology became a lot more interesting when Hubble found distant galaxies in the 1920s. Its size became vastly bigger seemingly overnight. However, Hubble was not the only scientist who was thinking about the cosmos. Albert Einstein (1879–1955) also had a monumental effect on our concept of it. In particular, his special and general theories of relativity rocked the comfortable world of Newtonian physics. It started this way: Einstein had been long fascinated with how we perceive things. He reasoned that in order to properly compare two events, we have to be able to measure them simultaneously. He often employed thought experiments to help him understand things. One of these thought experiments involved an observer at rest and an observer in motion both witnessing the same event, say the instant that a distant light is turned on. If the observer in motion is traveling toward the light's position at great speed, you would think he would observe the light before the observer at rest. However, there is overwhelming experimental evidence that contradicts that conclusion. In particular, the team of Michelson and Morley had conducted very precise experiments that found no difference in the speed of light due to motion through space.

Scientists of the world were baffled by the Michelson-Morley results, but one scientist saw the answer clearly. In 1905 Albert Einstein published his thoughts on this topic in the famous paper entitled the "Special Theory of Relativity." Einstein concluded that the speed of light is a constant and is accordingly the same value regardless of the conditions under which it is measured. In order for this to be true, time and distance must be variable. Wham! The world of Newtonian physics appeared to be turned upside down, but in a way it wasn't. The speed of light is very fast compared to ordinary experiences. Light can travel a distance equal to seven trips around the Earth at the equator in a mere second. And so, the ordinary, everyday measurements we make are unaffected. The variability of time and distance are only significant for travel at very high speed, like close to the speed of light.

How Einstein Revolutionized Cosmology

Albert Einstein roused a sleepy world with his general theory of relativity developed between 1907 and 1915. Sleepy because nothing really new had happened in the field of cosmology since Newton's time, and here was a radical new way of considering the universe: space and time were united as space-time, and the gravitational power of mass-energy shaped it. Moreover, Einstein provided the world with a mathematical expression to cement this relationship. Einstein's theory soon proved capable of solving problems not accounted for in Newton's gravitational physics. For one, it accurately predicted anomalies in Mercury's orbit. It also predicted gravitational waves, gravitational lensing, and gravitational time dilation.

Einstein also considered the nature of the cosmos. Is it finite or infinite? Is it stable, expanding, shrinking, or what? Several scientists had concerns regarding Newton's model of the universe because a slight imbalance in the distribution of stars could cause Newton's gravity to collapse the whole structure. Newton's model required an infinite universe to prevent that collapse. Einstein wrote that a finite universe could avoid collapse if space itself curved back on itself to form a sphere. At the time of these musings, Hubble had not yet made his discoveries that galaxies existed beyond the Milky Way and that the universe was expanding. Einstein assumed that it was stable, static, and had a uniform distribution of stars. However, his equations did not work using those assumptions. And that led him to make what he considered his greatest blunder. He added a fudge factor into his calculations in order to make the universe stable. He added that factor, designated by the Greek letter lambda, to his general relativity equation. It became known as the cosmological constant. When Hubble showed that the universe was not static but actually expanding, others derived the math needed to make sense of Hubble's discovery. These efforts led to today's view of an expanding universe, which began with a big bang explosion.

As for Einstein's cosmological constant, it has gone up and down in importance like a yo-yo. Einstein, once aware of the expanding universe, thought it had been a blunder to ever add it into the equation. However, in recent times, lambda has become a vital factor once again. "What changed?" you might ask. Well during the 1990s, astronomers found that not only is the universe expanding, but it is expanding at an accelerating rate. So, where does that driving force come from? It comes from a force, dubbed "dark energy." It is quite possible that Einstein's cosmological constant is dark energy. It is thought that dark energy can be envisioned as a constant amount of energy with negative pressure infusing all space. We will surely hear more about this topic in the future years.

Lead-Up to the Big Bang Theory

However, another model of the universe was going to challenge this Steady-State Model. The new theory proposes that the entire universe was once concentrated down to a singularity. In other words, to less than a pinpoint. Then one day, the universe burst forth with a bang. Sounds ridiculous, doesn't it? Others of the time thought so too. Out of derision, the Steady-State advocates named it the big bang theory, and the name stuck.

A good scientist strives to remain impartial and make judgments based on the facts. This is a worthy goal but not always easy to live up to. Yet, the facts supporting the big bang theory have won over the majority of cosmologists and astronomers. In fact, articles written in these fields reflect an assumption that the theory is true, and their arguments often build on that assumption. So, what are the facts that brought the big bang theory to prominence?

Edwin Hubble Expanded Our View of the Universe

We have already mentioned Hubble's discovery of galaxies outside the Milky Way, but the story needs to be told in more detail in order to appreciate it. Edwin Hubble (1889–1953) was fortunate in being able to do most of his astronomical work using the world's most powerful telescope at that time, namely, the telescope at the Mount Wilson Observatory. Moreover, he made that time pay off with monumental discoveries. Back in the 1920s, no one knew that galaxies existed outside the Milky Way. Faint-fuzzy objects were observed, but they were called nebulae and disregarded. There was also confusion as to how far away distant objects lie. Hubble put that problem to rest with perhaps the greatest discovery in the history of astronomy. He found a reliable measuring stick. He discovered that Cepheid variable stars could be used as yardsticks. These stars tend to pulsate radially and in doing so, produce changes in brightness with a consistent period and amplitude. In 1908 Henrietta Swan Leavitt had established these characteristics by observing thousands of variable stars in the Magellanic Clouds, which lie outside our Milky Way Galaxy. She showed that the true luminosity of a Cepheid variable can be determined by measuring its pulsation period. The distance to the star can be then calculated by comparing the known luminosity to the observed brightness.

Hubble put this technique to work establishing distances to Cepheids in the Andromeda Nebula. In doing so, he found the distances to be considerably outside the known size of the Milky Way Galaxy. He realized that galaxies of stars existed outside our local galaxy, and the universe became multiple times larger. Thereafter, the Andromeda Nebula became the Andromeda Galaxy.

Hubble Discovers an Expanding Universe

Among Hubble's notable achievements is his correlation between the redshift of light coming from a distant galaxy and its distance from us. In fact, his discovery is immortalized in his name, Hubble's law. Essentially, it says that the velocity that a galaxy is moving away from us is directly proportional to its distance from us. The constant of proportionality is Hubble's constant, which is a fundamental property of the universe. Hubble built upon the work of many scientists before him to reach this conclusion. One of them was Christian Doppler, who discovered the Doppler effect. We experience it most commonly with sound. The sound of a truck or train seems to change as it approaches us and passes us and then becomes more distant. Obviously, the sound that the truck or train makes remains a constant pitch. What is changing is the way we hear the sound waves emanating from something approaching us or fading from us. Light also behaves as a wave phenomenon, and it exhibits a Doppler effect too. Spectral lines from a receding object are shifted to the reddish end of the spectra and approaching objects toward the bluish end.

Hubble studied the spectral shift of forty-six galaxies and found with few exceptions that the galaxies are moving away from us. Moreover, as Hubble's law states, the farther they are from us, the faster they are moving away. Hubble calculated Hubble's constant from this data but arrived at a value that made the universe far too young. The distance values, he was using, were too far from the true values. More accurate distance values were obtained after the discovery of Type 1a supernovas and more powerful telescopes. Supernovas are exploding stars. Although a very rare event, when they happen in a distant galaxy, they light up with an intensity that is easily detectable despite the great distance. The Type 1a supernovas are very consistent in the energy they release and that makes them an excellent tool for determining astronomical distance. They are the standard candles of astronomy.

The Big Bang Theory Emerges

Georges LeMaitre

Georges LeMaitre (1894–1966) was the father of the big bang theory, which he called the Cosmic Egg Hypothesis. He was a man who wore many hats. For one, he was a Belgian Catholic priest. He taught college physics and pursued astronomy. Lemaitre was intrigued by Hubble's observations that the galaxies were moving apart from each other. LeMaitre felt that what was really happening was space was expanding, and the galaxies were carried along with the expansion. He mathematically analyzed this expansion in 1927 using the general relativity equations derived by Einstein. So, LeMaitre had actually anticipated Hubble's law, which states that the Doppler-shift-measured velocities of the receding galaxies are proportional to their distance for Earth. He also estimated that rate of expansion, called Hubble's constant. It was Hubble though, who did the pioneering astronomical work, which proved these relationships. LeMaitre is also credited with the concept of the universe beginning as an infinitesimal point.

Cosmic Microwave Background Radiation

The cosmic microwave background radiation, or CMBR, is one of the discoveries that proves that the big bang theory is valid. It is, in essence, the current temperature of the universe 13.7 billion years after the big bang. It has been measured to be 2.75 K, which is barely above absolute zero where all molecular motion ceases. The CMBR was accidently discovered by Arno Penzias and Robert Wilson in the 1964. They were awarded a Nobel Prize for their work on the CMBR in 1978. Cosmologists are keenly interested in how uniform the CMBR actually is because it gives them information about the early universe. Satellite studies of CMBR anisotropy have been done and are being done. The results match well with theoretical expectations. The most famous of these was NASA's Cosmic Background Explorer Satellite, or COBE, that orbited in 1989–1996. NASA launched a second CMB space mission in 2001 called WMAP and obtained more accurate readings. The European Space Agency (ESA) launched a third mission, called Planck Surveyor, which functioned between 2009 and 2013. With even finer precision. ESA launched the Planck Cosmology Probe in 2013. They obtained an age for the universe of 13.80 years and an estimate that the universe contains 4.9 percent ordinary matter, 26.8 percent dark matter, and 68.3 percent dark energy.

A Strong Argument for the Big Bang Theory Being True

Particle physicists have modeled that first second of the big bang. There are an enormous number of particles in the current universe. So, running the expanding universe film backward, we have compressed all those particles into a single entity. That entity would have to be an extremely dense plasma with a temperature over ten billion kelvins. Allowing the clock to run forward again, the physicists calculated the rates of nuclear reactions between protons and neutrons to form the elements out of the hot hydrogen plasma of the new universe. Lone unreacted protons would be the most numerous particles. They are the nucleus of the element hydrogen. Deuterium would be next most common. Deuterium is simply a rare isotope of hydrogen, containing a neutron in the nucleus along with the proton. The element helium would be the third most common. Helium is the combination of two protons and two neutrons. The next element lithium, which contains three protons, would be extremely rare. None of the other element would be formed in the big bang according to their calculations.

Now, if we look at the distribution of hydrogen and helium in the universe, that predicted distribution is exactly what we see. It is certainly true for our solar system. The sun is many times larger than all the planets and is mainly hydrogen and a little helium. The gas planets have a hydrogen to helium ratio much like the sun. Moreover, all the main sequence stars in the universe are mainly hydrogen just like our sun. This evidence won a lot of scientists over to the big bang side of the debate.

At this point, you may say, "Hey! Wait a minute. What about all the other elements?" Scientists are convinced that those elements did not exist in the early universe and that they had to be made in the nuclear factories of the stars. Main sequence stars convert hydrogen into helium by a process of nuclear fusion, but once the hydrogen is all consumed, new fusion reactions begin using helium. This process of creating heavier and heavier elements occurs faster and faster until the star explodes and flings these new elements into the surrounding space. They get taken up in the hydrogen clouds that form new stars and so on.

Time Line for the Big Bang Event

Scientists have calculated the time that the big bang event occurred by running the clock backward. We know how fast the galaxies are moving away from each other, so reversing the process brings us to time zero when the big bang explosion occurred. That is 13.8 billion years ago. The following table is a time line for the series of events that followed.

Table 1.1 Time Line for the Big Bang Event

Time	Event
10^{-43} seconds	The Big Bang
10^{-6} seconds	The universe takes shape
3 seconds	The elements hydrogen, helium, and lithium form out of plasma
10,000 years	The Radiation Era
300,000 years	The Matter Era
300 million years	Formation of stars and galaxies
8.7 billion years	Our sun and solar system are formed

As the table shows us, a lot happened in the first second. Enormous heat was released in the process. The universe expanded in a millionth of a second in a process dubbed "inflation." Cosmologists are still at work conducting experiments to confirm their theories of this mind-numbing inflation event. That universe was a seething soup of electrons, quarks, and other particles. The expansion process caused cooling and at a certain point, the quarks clumped into protons and neutrons. The universe was a superhot fog and essentially opaque at this point. By three hundred thousand years old, the universe was cool enough for atoms to form, and this process led to the transparent universe we have today. When the universe was three hundred million years old, it had cooled enough for clouds of hydrogen to form and the first stars to ignite. Many astronomers around the world are focused on enhancing their telescopes to see back to the first stars and galaxies. This will be an exciting field to follow over the next decade or two. As these early stars eventually went supernova and exploded, they created the heavier elements necessary for us to exist.

Research to Help Us Understanc the Big Bang Continues

The Earliest Galaxies

Telescopes are in a sense a time machine; they tell us about past events. If you just saw the sun in the sky, what you really saw was the sun as it was eight minutes previous. It took that long for the light to travel from the sun to us. Astronomers can look backward in time over thirteen billion years to witness some of the earliest galaxies. Moreover, there is a lot of interest in seeing these first galaxies because this information helps scientists better understand the big bang chronology. The James Webb Space Telescope (JWST) will help astronomers see further back in time than they previously were able. It is being jointly developed by NASA and the European Space Agency and the Canadian Space Agency. It is somewhat like a modernized Hubble Space Telescope, although it is different in the sense of seeing clearly in the infrared. Very distant objects have their light shifted into this part of the spectrum. The satellite must be shielded from infrared heat sources that would interfere. The JWST has an anticipated launch date of October 2018.

Even the best telescopes cannot see farther back than to a time three hundred million years after the big bang event because the universe was opaque for a period known as the Cosmic Dark Ages. And that is where supercomputers come into the picture. What we cannot observe, we model. If our computer model predicts things that we have observed or can observe, then we believe we have the correct model. When scientists modeled the universe using the mass of visible objects like stars for the mass of the universe, the model failed badly. However, when they added additional mass to their value, the model worked. This mysterious missing mass has been given the name "dark matter," and it is a subject of great importance in physics. Scientists are working on understanding dark matter in institutions around the world.

How Matter Was Created

Scientists are intensely interested in that first instant of the big bang event. What exactly was created in that first instant? Would finding that out be an impossible task? Not to the scientists at Brookhaven National Laboratory (BNL) in Upton, New York. This research laboratory is run by the US Department of Energy. Science now accepts the concept that the matter of the universe was formed out of a hot plasma starting 380,000 years after the big bang and continuing for another 100 million years. In order to simulate the big bang event, BNL scientists are smashing super-fast-moving gold nuclei together in their supercollider. The gold nuclei are traveling at close to the speed of light when they collide. The result is a super energetic event with hundreds of particles being formed.

Cosmic Microwave Background Radiation

Earlier in this chapter, a chronology of CMB experiments was mentioned. These experiments are extremely important to big bang theorists because the fine structure of the early universe can determine which theoretical assumptions are correct. The latest of the CMB experiments is called the PiPER Telescope, and its goal is to learn more about a big bang event called "cosmological inflation." This refers to a Nano-Nano-nanosecond period when the early universe expanded at a speed faster than the speed of light. In a NASA-funded project, the Primordial Inflation Polarization Explorer (PiPER) will be lifted by balloon to a height of twenty-three miles, where it will conduct measurements. NASA hopes to do multiple flights with the telescope. PiPER may provide a linkage between gravity and quantum mechanics.

Stars

The theory underlying the big bang expansion tells us that the expanding embryonic universe was cool enough for hydrogen atoms to form after about four hundred thousand years in an event known as "recombination." Until that time, the universe had been a white-hot plasma of protons, neutrons, and electrons. The recombination is significant because it caused the universe to go dark and stay dark for millions of years. The first stars formed around one hundred million years after the big bang. The existence of hydrogen made those stars possible. Remember that stars (and our sun is a star) are actually thermonuclear reactors, where hydrogen gas has been compressed to the point where a fusion reaction begins at a temperature of fourteen million kelvin. However, these first stars were markedly different from the stars that we observe today. They were immensely bigger, perhaps a hundred times more massive than our sun. Although it may seem counterintuitive, bigger stars have shorter lifetimes than smaller stars. So, it follows that these first stars had very short lifetimes compared to stars like our sun. Typically, stars die by running out of nuclear fuel and exploding, but that is an oversimplification. When hydrogen runs out, a new nuclear reaction begins, which uses helium as a fuel. The products of this reaction are carbon and oxygen. Depending on the mass of the stars, different series of reactions occur. In general, this nucleosynthesis process accounts for the rest of the elements not manufactured during the big bang event.

In the final death of a star, it typically explodes and expels these new elements into the surrounding space. When these regions of space are later compressed to create new stars, they are mainly hydrogen but also include the dust from previously exploded stars. This is how heavy elements become incorporated in new stars that are forming. Did you know that stars are being born right now? Look up in the evening sky and spot the Pleiades or Seven Sisters constellation. It is a nursery for new-forming stars. Or find Orion in the winter sky and examine his belt with binoculars. It is another star factory.

Our sun is 4.6 billion years old, which means the universe was about nine billion years old when it was born. There had been sufficient time for many generations of stars to form and die during that time. Each generation enriched the content of heavier elements. The planet that we live on is composed of those elements that were created in previous stars. We ourselves are made of stardust.

Astronomers can measure the metallicity of a star from its spectrum. Here the word "metallicity" refers to the elemental content of the star heavier than helium. Yes, they can tell the elemental makeup of a star from its electromagnetic radiation using spectroscopic analysis. The fact that they consider elements heavier than helium to be "metals" is a bit odd, but then star environments are so hot that normal chemical reactions are nonexistent. Stars with high amounts of carbon, nitrogen, and oxygen are deemed to be "metal rich" by astronomers. Earth-bound chemists do not think of them as metals at all. A useful aspect of metallicity is to gauge the age of stars. In 1944 the term "Population I" was invented to describe metal-rich stars and the term "Population II" for metal-poor stars. In 1978 the term "Population III" was invented to describe extremely metal-poor stars. The first stars formed one hundred million years after the big bang are Population III stars.

Now back to the dark universe after recombination. Theory also tells us that the first stars had little effect on brightening the dark universe. The reasoning is that these stars were brightest in the ultraviolet portion of the spectrum and that frequency range was absorbed by the hydrogen gas in the universe. And yet the dark period did eventually end. By the time the universe was one billion years old, the dark period was no longer. Physicists refer to the phase that made this possible as "reionization" and are currently attempting to understand the processes involved. Progress is being made. The most helpful information comes from astronomical observations, which reach far back in time. The most distant stars are the oldest stars in the universe. The light from them has been traveling to us for many billions of years. How can we see such dim stars? Here is one example: The Hubble Space Telescope was trained on one spot in the sky for endless hours in order to accumulate enough ancient photons to see back to thirteen billion years ago. This is within eight hundred million years of the big bang event. Newer telescopes are coming into operation in the years ahead with the power to look even farther back in time (Source: Sci. Am. April 2014).

The Milky Way Galaxy

The first galaxies were being created during that dark period of the early universe, and one of them may have been our own galaxy, the Milky Way Galaxy. Have you ever seen it? I grew up in a rural area of Illinois, free of city lights. Our night skies were often perfect for observing the bright stars of the constellations, but most of all, I was impressed by the sweep of dense Milky Way stars, which crossed the entire sky. Many city-raised kids today have never seen it and wouldn't know what it was if they saw it. Galaxies come in a variety of sizes and shapes. Our galaxy is what is called a barred spiral, and it has a diameter of 100,000 to 180,000 light years. In other words, it would take you a minimum of one hundred thousand years to transverse it if you could travel at the speed of light, and Einstein said it is impossible to go that fast. Our galaxy has a huge number of stars in it. Estimates range from one hundred billion to four hundred billion of them. One of those stars is our sun.

Figure 1.1 Milky Way Galaxy
By 4_Milky_Way_(ELitU).png: Andrew Z. Colvin derivative work: Frédéric MICHEL (4_Milky_Way_(ELitU).png) [CC BY-SA 3.0 (http://creativecommons.org/licenses/by-sa/3.0) or GFDL (http://www.gnu.org/copyleft/fdl.html)], via Wikimedia Commons.

Our solar system lies twenty-six thousand light years from the galactic center on the Orion Arm. If you wonder where the galactic center is, look toward the summer constellation, Sagittarius. I usually look for the constellation Scorpio, which looks like a scorpion, and Sagittarius is close by. As the galaxy rotates, our solar system makes one rotation every 240 million years. Although our solar system seems huge to us, it is miniscule compared to the size of the galaxy. Let's discuss our solar system next.

Our Solar System

There are two very different kinds of planets in our solar system: rocky planets and gas planets. The rocky planets are Mercury, Venus, Earth, and Mars. The asteroid belt lies between Mars and Jupiter and may have failed to become a planet due to Jupiter's disruptive influence. The gas planets are Jupiter, Saturn, Uranus, and Neptune. They are composed mainly of hydrogen and some helium. Our Sun has a similar composition. It is noteworthy that these two elements are the major elements thought to have been created in the big bang. Pluto, outermost and composed of ice, used to be counted as a planet but has been demoted to dwarf-planet status. Several other dwarf planets exist (Haumea, Makemake, and Eris).

Early Solar System

Jupiter is the largest planet in the solar system and has two and a half times more mass than all the other planets in the solar system. The position of the planets has long been considered fixed, but that view has changed based on what has been learned about exoplanets. Jupiter-size planets have been detected orbiting close to their stars and are called "Hot Jupiters." It is impossible that they formed where they currently are, so they must have drifted there after forming much farther out from their stars. This new information makes us reconsider the stability of our own solar system. It is now thought that Jupiter was one of the first planets formed, and it drifted inward toward the Sun. It might have spiraled into the Sun had it not been for Saturn, the next most massive planet. The gravitational attraction between Jupiter and Saturn caused them to adopt the orbital relationship that they have today. This theory helps resolve two other anomalies: (1) Why Mars is much less massive than Earth and Venus and (2) why the asteroid belt contains both rocky and icy bodies. Mars was deprived of building material by the inward journey of Jupiter, and the sweeping action of the planet caused the current composition of the asteroid belt.

Planet Number 9

Our farthest planet Pluto has been a mainstay throughout most of my life. However, in August 2006, it was ignominiously demoted to dwarf-planet status. I have to tell you that it feels a little strange being reduced to only eight planets, when there were nine planets through most of my life. So it is pleasant to learn that there may be a ninth planet in the line up again. The problem is, that of this writing, astronomers do not know where that planet is exactly. They do know that it is somewhere in the Kuiper Belt, that zone outside Neptune where all the icy bodies reside. They also know that it is ten times more massive than the Earth based on its disruptive influence on the orbits of the icy bodies in its vicinity. It seems unlikely that it is a rocky planet or an icy planet based on its size and the sparse amount of material in the Kuiper Belt from which to construct it. It is more likely that it is a Mini-Neptune (i.e., a gas giantlike Neptune, only smaller). Astronomers have not ruled out the possibility that it is an alien planet captured from another star. I can't wait to learn more about this ninth planet in the future.

Formation of the Earth/Moon System

Our Position among the Rocky Planets

There are four rocky planets orbiting close to the Sun and four gaseous planets orbiting farther out. There might have been a fifth rocky planet, but its formation failed, and we have the asteroid belt in that orbit instead. So what can we say about those rocky planets? Mercury is closest in at 36 million miles from the Sun, and that is far too close for life to exist. One face of Mercury always faces the Sun, and metal will melt on that surface. Venus is the next planet out from the Sun at a distance of 67 million miles, and it is also too hot for life as we know it to survive. That is unfortunate because Venus is very similar in size to the Earth and had it been cooler, might have been another potential home for us. Runaway global warming is to blame for her sizzling climate. Our planet Earth is next out from the Sun, but let's save it for last. Mars is the fourth planet out from the Sun at some 142 million miles. The Red Planet has been a favorite for speculating on alien life in solar system. In the early days of telescopes, some thought they spotted canals on the surface. Today, we know a lot more. Scientific missions like the "Curiosity" rover have moved around of the surface of Mars and analyzed the soil to determine if the planet once had life. It is still an open question.

Our Planet Earth

Figure 1.2 Our Planet Viewed from Space
By NASA Goddard Space Flight Center—http://www.flickr.com/photos/gsfc/6049973495, CC BY 2.0, https://commons.wikimedia.org/w/index.php?curid=37986108.

And that brings us to Earth, our home planet. We orbit the Sun at a distance of ninety-three million miles, which seems to be the sweet spot climate-wise. Like the other planets, Earth was constructed of smaller bodies called planetesimals. The process began during the Sun's birth some 4.6 billion years ago. Dust particles in a vacuum are drawn together. These dust clumps collided and became bigger clumps. The process continued as bigger and bigger bodies formed. This early period was a time of heavy bombardment as we can see looking at our moon through a backyard telescope.

Formation of Our Moon

I have watched the scientific community go through a series of refinements in their theory of moon formation. Back in the twentieth century, several different hypotheses coexisted. However, the community focused tighter and tighter on the impact theory, where a Mars-size protoplanet, name Theia, hit the Earth at a glancing blow. The resulting debris formed a disk, which initially circled close to the Earth. While some debris fell back to Earth, the rest coalesced into a new body that became the moon. The Moon has been receding from the Earth ever since.

The latest position on Moon formation at NASA Lunar Science Institute is that the Earth and Moon were created together in a giant collision of two planetary bodies that were each five times the size of Mars. The new hypothesis is based upon modeling experiments plus the fact that the Earth and Moon have the same composition. This would not have been the case with the previous concept.

First Life on Earth

The newly formed Earth and Moon had molten surfaces for millions of years. It was too hot on either of them for life to form. Life never did form on the moon. It is too small to hold an atmosphere, it offers no protection to harmful radiation, and it offers no protection against very high velocity particles. Life did form on the Earth as soon as it sufficiently cooled. That will be the topic of the following chapter.

Two

Early Earth, Oxygen, and the Evolution of Multicelled Life

The Chapter in a Nutshell

As the planets of our solar system formed from planetesimals, it was a time of heavy bombardment. The pockmarked face of the moon gives us an indication of what it was on Earth. The surface temperature of the Earth was too hot for life to exist and remained that way for hundreds of millions of years. When the planet finally cooled enough for life to exist, there was insufficient oxygen in the oceans or in the atmosphere for multicellular life to form. Single-cell life did get started though, and cyanobacteria evolved to use photosynthesis to build organic molecules. Most importantly, it expelled oxygen as a waste product. Over time, the oxygen content of the oceans and atmosphere gradually increased, but over three billion years elapsed before the oxygen content was high enough to sustain multicelled organisms.

Eventually multicellular life burst forth! It is a geological principle that deeper strata are older strata. The deepest strata containing visible animal fossils were dated at 540 million years old, and the number and diversity of these seafloor life-forms were surprising. It seemed to be an explosion of life, and so it was dubbed the Cambrian explosion. While the discovery was a fortuitous for science, it became was a worrisome problem for Charles Darwin and his theory of evolution. Darwin taught us that living things change to very gradually adapt to changing environments. That evolutionary adaptation happens over numerous generations. Moreover, we should see fewer, simpler life-forms precede more numerous, more complex life-forms. And yet the Cambrian explosion seemed at first to disprove that theoretical prediction. Eventually, Darwin was vindicated when microscopic multicelled organisms were detected in lower strata.

One of the frontier areas of the life sciences is Evo Devo, which is short for evolutionary developmental biology. This is a field of biological research concerned with how different life-forms got to be the way they are and how the processes involved actually accomplish these transformations. Evo Devo combines embryology with DNA and genetic science to reveal how nature builds animals from a fertilized egg. The diversity of Cambrian creatures can be understood as variations on a theme of segmented body parts. Regulatory genes, particularly Hox genes, control the development of different body segments.

Oxygen and Multicelled Life

This chapter is about the newly formed planet earth and how life formed on it, and yet the first thing we discuss is the element oxygen. You will see in the discussion to follow how multicellular life was dependent upon the presence of oxygen to survive on earth.

Oxygen, a Vital Element

What is your favorite element? That is probably a question, which you are not often asked. Yet, if you think about very long, the element oxygen is the most sensible answer. If we could not draw adequate oxygen into our lungs in the next few seconds, our lives would be in great jeopardy. The air around us consists of four-fifths nitrogen and one-fifth oxygen, but that high level of oxygen is a fairly recent development when you think about oxygen levels during the entire age of the planet. In fact, we will go back even further to the time when planets didn't yet exist and a ring of debris defined the material around our young sun.

Scientists wondered about the process by which planets formed in those early days of the solar system. The answer in part came from experiments conducted by scientists in space. They learned an important fact that dust and fine particles automatically clump together in the vacuum and weightlessness of space. Small clumps become bigger clumps, and the snowballing process continues until planet-size objects have formed. Then gravity exerts its influence, pulling asteroid-size bodies toward each other. We can now envision what it was like in the early days of the solar system: there were hundreds of large objects (i.e., planetoids) orbiting the young sun, and violent collisions frequently occurred between them. Just look up at our moon on a clear night with binoculars, and you will see proof of such impacts. Now, some of these impacts demolished the planetoids. Other times the impacts ejected one of the bodies from the solar system, and sometimes the collisions resulted in creating an even larger body. This is how the planets formed, and the earth was one of those planets.

After the Earth was formed from colliding planetoids, the collisions continued for many millions of years. However, the material available for collisions diminished with time and finally slowed to a near stop. Until that happened, the planet was far too hot to retain an atmosphere. However, even when it finally cooled enough to retain an atmosphere, oxygen was barely present. The gas known as oxygen (i.e., O_2) is only formed in two ways: one is by photosynthesis, and the other is the breaking apart of water molecules by cosmic radiation. Both hydrogen and oxygen are generated by this latter process, but hydrogen, being lighter, quickly escapes from the Earth.

Would you think that oxygen so vital to our lives, is dangerous to our lives as well? Wouldn't you think that if a little oxygen is life sustaining, then more oxygen is better still? Unfortunately, that is not the case. We long ago learned the danger in giving divers pure oxygen, that below a depth of twenty-six feet they would experience grand mal seizures and lose consciousness. Well, okay, pure oxygen is obviously dangerous, but surely lower levels are safe. Sorry, we cannot conclude that either. Animal studies at 75 percent oxygen levels resulted in inflammation of the lungs within a few days followed by death. The reason that oxygen is both a necessary and a dangerous substance has to do with free radicals. Most people have heard the term "free radical" before, usually in discussions about aging. The advice that usually accompanies these aging discussions is to take more antioxidants into our bodies to combat free radicals. It is not that simple, but that discussion is for another day. For now, it is enough to know that breathing oxygen, that is, respiration, is simply a much slower form of combustion in the same way that rusting is a slower form of combustion. Hundreds of millions of years of evolution are responsible for our adaptation to an oxygen-breathing lifestyle. It is a highly efficient way of using energy, but it has its long-term costs as well.

Oxygen and Living Things

The Early Earth

We are less interested in the chemistry of oxygen and more interested in life on Earth, but it turns out that the two are strongly intertwined. One of the authors who has written on this relationship is Nick Lane. His book *Oxygen, the Molecule That Made the World* has been a source for this chapter. Another expert is Peter Ward, and oxygen plays a starring role in his two books *A New History of Life* and *Out of Thin Air*. Now, in order to understand why living things are the way they are or how they got to be the way they are, we need to understand the continual relationship between life and the availability of oxygen. To this end, let us time travel back to the early days of the solar system and see what the relationship was then. Actually, there is none yet because both life and oxygen were missing from the early Earth. The young planet Earth was simply too hot for either of them to exist. When the bombardment period finally subsided and the molten Earth cooled enough to sustain an atmosphere, there was almost no oxygen in it. There was no life either for half a billion years. When life finally appeared, it was primitive and single celled. In fact, it was another three billion years before multicelled life-forms are seen in the fossil record. Why the long delay? Oxygen had to reach a high enough percentage in the atmosphere for multicelled life to exist.

So how did oxygen gradually build up in the atmosphere, lakes, and oceans? The main process of oxygen generation is photosynthesis. One of the earliest life-forms, capable of using it, is cyanobacteria. This bacterium found a way to utilize the energy of sunlight to incorporate carbon dioxide from the early atmosphere into organic compounds. In the process, it expelled oxygen as a by-product of the reaction, and oxygen was toxic to many of the organisms of the time. So, although oxygen is vital to multicelled life, early organisms had to evolve defenses against its toxic effects. One of the defenses was for cells to clump together. This defensive strategy may have helped drive multicelled organisms to evolve in the first place. Single-celled organisms, such as bacteria, do not need oxygen for generating energy. However, multicelled animals must use oxygen for energy and that is because it takes a lot more energy to move around than to staying stationary. Aerobic respiration, that is, breathing oxygen, yields ten times more energy than the anaerobic respiration, which bacteria use.

We tend to think of living, moving things as existing on land, but that is not how life evolved on Earth. The first multicelled animals lived in the oceans and lakes of the world. So, let us consider the differences between breathing air and breathing water: There are thousands of fewer oxygen molecules in a given volume of air than there are in the same volume of water. That is why lungs for air breathers are much bigger than gills for water breathers. Moreover, if you are an air breather, not all air is the same. The higher the altitude at which you live, the less oxygen is in the same volume of air. By the same token, for water breathers, not all water is the same. You will get more oxygen from cold water than from warm water, and freshwater lakes and rivers hold more oxygen than do saline oceans. Then there is the factor of how hard one works to process seawater for oxygen. Those passive sea creatures tend to be sluggish and stationary. They have even minimized their nervous tissue because so little oxygen being obtained. Sea animals, which are active and aware of their surroundings, must effectively extract the most oxygen they can from the water. So, how does the oxygen get from the lungs or gills to the cells, which need it to function? Of course, the answer is by means of blood circulation, but that raises the question of how does the blood do the job. It turns out that different animals use different oxygen-carrying chemicals to do it, although the principle is the same. We humans, all vertebrates, and many other animals use hemoglobin. This protein has an atom of iron in it, which makes the transfer of the oxygen to the cell possible. However, other animals like crustaceans, use hemocyanin, which has a copper atom at its center. A third category use hemerythrin, which is also iron based.

LUCA and Relatedness of Living Things

Even before cyanobacteria evolved on the Earth, there were earlier life-forms. One of them was a common ancestor of all life on our planet. Yes, science has established that all life that you see around us is related to each other through a common ancestor. LUCA is shorthand for the Last Universal Common Ancestor. In other words, LUCA ties all living things together. So, what was LUCA like? We don't have an exact answer to that question, but we can infer LUCA's identity from its offspring that exist today. There are three great domains of life-forms, namely: bacteria, archaea, and eukaryotes. While I am sure that you have a feel for what bacteria is, you may not be familiar with archaea or eukaryotes. Let's consider eukaryotes first because we belong in that domain and so do most of the things that we love. As you may know, our bodies consist of trillions of cells. Now, inside our cells lies a nucleus, which houses our DNA. Having that nucleus is what makes us eukaryotes. All multicelled life have that nucleus and are also eukaryotes. However, the first life-forms did not have a nucleus and are consequently called "prokaryotes." It turns out that the other two domains (i.e., bacteria and archaea) are prokaryotes.

The last domain to define are the archaea. Besides being prokaryotes, what can we say about them? Some would say they are the missing link between multicelled life (i.e., eukaryotes like us) and bacteria. However, another description of archaea is that they are extremophiles. They live in the harshest environments, such as boiling-water temperatures or frigid temperatures, acid conditions, caustic conditions, or saline conditions. They are also nonpathogenic, which means that they do not invade us as do bacteria.

The Cambrian Explosion

Single-celled life began with LUCA, evolved into bacteria, archaea, and eukaryotes but failed to make the leap to multicellular life for billions of years. Incidentally, only eukaryotes were able to become multicellular life. However, when multicellular life made its debut in the fossil record, it did it with a bang. That is our next topic.

The Burgess Shale

The cyanobacteria, using photosynthesis, gradually transformed our atmosphere from one lacking oxygen in its makeup to one having significant oxygen. How do we know this? The lower strata of the Earth contain iron in its elemental state, whereas all strata above these levels contain iron as an oxide. Thus, we have the red-tinted hills, which make western movie scenes so colorful. Of course, the oxygen content of the atmosphere increased very gradually over many hundreds of millions of years. Multicelled life had to wait for eons to evolve because there was inadequate oxygen for them to thrive until about eight hundred million years ago.

The earliest strata containing visible fossils have been dated as 530 million years old. Any strata much older than this seemed to be devoid of visible life. Recent work shows that multicellular life was present earlier, but it took digging to see it. There is no doubt about the fossils found 530 million years ago. These strange-looking, ocean-bottom creatures belong to the Cambrian Period, and they still fill scientists with awe. The Cambrian is also important because it is the first period in the Paleozoic era. We will be discussing that era in the next chapter.

Now, just what kind of creatures lived in the Cambrian might never have been revealed to us were it not for an unusual event that happened over half a billion years ago. A massive mudslide in the ocean carried a group of these early creatures down into a toxic zone lower in the ocean and killed them in their prime. Over time, that ocean bottom became sedimentary rock, known as limestone shale. This particular shale region, containing the Cambrian fossils, is called the Burgess Shale and is found in the mountains of Alberta and British Columbia, about a ninety-minute drive from Calgary. I can tell you from personal experience that the Canadian Rockies are a most beautiful place to visit. However, I never got to see the Burgess shale site because its access is restricted to the public. Most of the Cambrian fossils are stored in the Smithsonian Museum, anyway.

Charles Doolittle Walcott

The person who retrieved, researched, and published his findings on these astounding animals is Charles Doolittle Walcott (1850–1927). At the young age of twenty-two, Walcott partnered with William and Hiram Rust in excavating the Walcott-Rust quarry for trilobite fossils. This site was an important discovery because these beautiful fossils had clear imprints of appendages and soft tissue. Incidentally, Walcott fell in love with and married Laura Ann Rust, but more significant for his career as a paleontologist, Walcott became somewhat of an expert in trilobites and published several papers on them. Walcott ended up selling the trilobite collection to the Museum of Comparative Zoology, Harvard University.

For a man who never finished high school, Walcott did a lot with his life. His interest in fossils landed him a job as assistant to paleontologist James Hall, and from there to a position with the US Geological Survey in 1879. By 1894, he was its director. He was elected to the Academy of Sciences in 1896 and was one of the founders of the Carnegie Institution of Washington. By 1907 he was secretary of the Smithsonian Institution, a post he held for the rest of his life.

Walcott had an impressive career, but the discovery that makes him a historic figure is his discovery of the Burgess shale and the fossils of the bizarre creatures that had once inhabited it. Between 1910 and 1924, Walcott made numerous trips to this fossil treasure trove and returned to the Smithsonian laden with fossils. In all, he collected sixty-five thousand specimens. As he found time while also attending to his administrative duties, to examine and report on these fossils from the earliest time in which we see multicelled life.

Meet the Creatures of the Burgess Shale

What astounds most people when they see the creatures of the Burgess shale is the diversity of differing life-forms from such an early time. No wonder this fossil array is referred to as the Cambrian explosion. Diverse life-forms seem to have exploded upon the scene out of nowhere.

I don't see how it is possible to capture the excitement that scientists experienced from the Cambrian menagerie, if you have never seen these bizarre creatures. Of course, you could visit the Smithsonian museum, where they are housed and exhibited. However, second best is for me to show you some pictures of these bizarre animals artistically brought back to life. These creatures lived on the ocean bottom and therefore had a way of extracting oxygen from the seawater. Often they had legs that could both propel them and also serve as gills.

The images below all come from the website Wikimedia Commons. These are but a small sampling of the animals living on the ocean bottom from 530 million years ago. If you would like to learn about these and other Cambrian creatures in greater detail, I recommend that you get a copy of Stephen Jay Gould's book, *Wonderful Life*.

Figure 2.1 Marine Animals of the Cambrian

Courtesy of Wikimedia Commons

Aysheaia

Aysheaia was a cylindrical-shaped animal about two inches long, having ten unjointed legs. The head was not distinct from the body and had several spikes on it as seen in the photo. Its mouth was in the middle of those spikes. The Aysheaia fossils were found with sponges, so it may have liked to eat sponges or may have just hidden from predators among them.

Waptia

Waptia was a three-inch-long arthropod with a carapace head shield that hinged in the middle. Beneath the shield was a head with a pair of long antenna and a pair of eyes on stalks. The long slender body had ten pairs of appendages. The abdomen consists of five elongated segments, and it ends in a forked tail with oval blades. Waptia was a common find in the Burgess shale. Over fourteen hundred specimens were collected.

Yohoia

Yohoia was an arthropod, something like a trilobite. It has a head shield and thirteen rows of tergites. A tergite is a trunk plate. Unlike trilobites, it had a paddle-like tail and two arm-like extensions in front of the head. Yohoia roamed the ocean bottom either scavenging or hunting for prey.

Opabinia

Opabinia is a rare fossil; only twenty good specimens have been found. It doesn't fall into any of the known phyla but is thought to be descended from arthropods and annelid worms. They ranged between 1.6 and 2.8 inches in length and were so weird looking that people seeing their images laughed. It had a hollow proboscis that was one-third of the body length. Apparently, it has used like a vacuum cleaner hose to suck up morsels. It had five eyes, some of them on stalks. The body consisted of fifteen segments and ended in a V-shaped double tail.

Anomalocaris

Anomalocaris was huge compared to the creatures discussed so far. Their bodies could be over three-feet long. This giant is thought to have been a predator that could swim above the sea bottom. It had eleven lobes, which served as fins to allow it to swim. It had a large head with a pair of compound eyes on stalks. The eyes had a total of about sixteen thousand lenses. It had a circular mouth below with the power to crush its prey. In front of that ominous mouth were two large arms with barb-like spikes.

Wiwaxia

Wiwaxia looks like a plant but is really an animal. Some think that it resembles a porcupine. They measured up to two inches in diameter. They had spines arranged in two rows, which served as protection from predators. Otherwise the upper body was covered with small plates. They had a mouth and are thought to be a bottom-feeder.

Pikaia

Pikaia was a two-inch-long, symmetrical creature having a head distinct from its tail and two forward-looking eyes. Of all these Cambrian creatures, Pikaia may be the one most likely to be our distant ancestor. It is a chordate. In other words, it has a primitive spine called a notochord. All animals having a spine are chordates as a biological grouping. Vertebrates, like us, are a grouping under chordates.

A Three-Dimensional Viewpoint

The authoritative source on the fossils of the Burgess Shale is the late Stephen Jay Gould, and his famous book on it is titled *Wonderful Life*. Gould tells us that Walcott did his best characterizing the creatures, which he discovered, but was too weighed down by administrative duties to do the characterizations using the best techniques. Walcott seemed to regard these fossils as flattened impressions of what they had been when they were alive. He also was careless about preserving both halves of the shale after it was split to reveal the fossil. Decades later, the Cambrian fossils were reexamined with new results. It took the skill level of Cambridge geology professor and trilobite expert, Harry Whittington, to see them as three-dimensional creatures and to analyze the fossils anew from that perspective. Whittington and his assistants took advantage of the multiple orientations, which the fossils of a particular animal can be found in to recreate the animal as it was before being flattened. They also searched for both sides of the fossil-containing shale in the Smithsonian's storage rooms because the missing half sometimes contained some of the animal on the reverse side. Delicate slicing of the fossils allowed them to examine in a series of layers. They were able to reconstruct three-dimensional drawings of the animals showing carapace, gills, and legs in great detail.

Charles Darwin and the Cambrian Explosion

An Idea That Shocked the World

Imagine yourself living in England back in 1859. Charles Darwin (1809–1882) had been secretly working on his theory of evolution for over twenty years, and circumstances are forcing him to either publish it or see the credit for his theory go to someone else. The dominant thinking in Darwin's day was that species were immutable. In other words, the species were created by God in perfection and therefore would never change. Religious doctrine had dominated such topics for centuries, and few folks were brave enough to challenge that authority. The problem with the immutability concept is that the fossil record was telling a quite different story. Geology was getting established as a scientific field in Darwin's time, and it was realized that deeper strata were older than shallower strata. Different kinds of animal fossils were associated with these different strata. Fossils of long-extinct animals, like massive dinosaurs, when compared with fossils of today's animals, clearly show that an animal form is not invariable.

Nature's Laboratory

Darwin was not relying on speculation as he formulated his theory. He had studied natural science in the laboratory of the world. In 1831, as a young man, Darwin put his shrewd observational powers to work as he witnessed plants and wildlife in South America. He sailed with Captain Robert Fitz-Roy on the HMS Beagle for five years as a naturalist and personally saw evolution at work. Although their voyage took them around the periphery of South America, and remarkable things were witnessed on outings there, it was most dramatically in the Galapagos Islands, where Darwin saw profound evidence of evolution. These islands were inhabited by finches as well as other animals, but the finches exhibited traits that were unique to the island they inhabited. Darwin noted that the vegetation also varied island to island and that the finches had adapted to best utilize their available foods and climates. Darwin postulated that the fittest animals survive and pass their superior traits forward to their offspring, whereas the lesser-adapted animals tend to die off. He called this culling process "natural selection." In other words, nature is deciding who lives and reproduces or who dies before they are old enough to mate. Over many generations, natural selection will change a species to better adapt to its new environment.

Just so you know, Darwin also discussed another kind of selection with which we are familiar: Since the earliest days of agriculture, man had been making such "live-or-die" decisions about grains or livestock. For example, beef cattle are radically different from dairy cattle. This is the result of many generations of man selecting which of their animals get to mate in order to optimize the cattle for their end purpose (meat or milk). Darwin called this process "artificial selection." The hundreds of different breeds of dogs, which exist today, have all been tailor-bred using artificial selection. In chapter 9, we will see this human innovation change humankind and the world.

Natural Selection, Codiscovered

Although Darwin's theory of evolution was gelling in his mind as early as 1836, he waited until 1859 to publish it. That is over twenty years of hesitation. He did publish his observations on natural history during the Beagle's voyage, and he published his ideas on the structure of coral reefs. These publications brought him widespread fame and respect. And yet, he had not published his ideas on evolution that were a major achievement. He might not have published these ideas even in 1859, if it were not for Alfred Russel Wallace. This naturalist went through the same kinds of experiences and the same kinds of reasoning processes as Darwin, except his islands were in the Malay Archipelago. Wallace had codiscovered natural selection and had approached Darwin for advice regarding the publishing of his discoveries. Darwin, in turn, consulted with his close friends, and they advised him to quickly publish or lose credit for what was his greatest insight. And that led to the bombshell publication that rocked the world. Wallace was acknowledged for his contribution in the process. That all happened when Charles Darwin shocked the world with his book *The Origin of Species*.

Darwin's Dilemma Resolved

When the Cambrian fossils were first found, it posed a serious problem for Darwin's theory of evolution because these first creatures were too diverse and too complex to have been the first animals. Darwin's theory required that evolution occur in a slow, gradual way and that simpler forms preceded complex forms. Any generation should seem to be indistinguishable from the previous one with changes very subtle and gradual. But the Burgess Shale told a different story; one of myriad forms bursting forth with no simpler forms preceding them. Darwin's dilemma was that he needed to explain the fact that the Precambrian strata seemed to be devoid of fossils. Darwin was troubled by it but believed in his theory enough to predict that eventually more primitive fossils would be found.

Darwin was right to have faith in his theory. At first, the puzzle seemed unsolvable. More scientists joined the quest, but the Precambrian creatures continued to elude the researchers. Then the Cambrian puzzle was finally resolved: Precambrian creatures were eventually found, and they were more primitive organisms. One of the clues was that after 800 million years ago, stromatolites are rarely found. Stromatolites are fossilized clumps of minerals and cyanobacteria. We now know that something had started eating the developing stromatolites. That something was one of the first multicelled animals. Plankton developed about then, and the earth oxygen content rose faster. Fossils of these earliest animals never were found because the animals were too small and had only soft parts. Then actual Precambrian fossils were found. One group is the Ediacaran fossils, found in Australian sandstones. They were reminiscent of coral, jellyfish, and worms. Some experts think these creatures were not ancestors to the Burgess Shale animals. Then another Precambrian group was found, the Tommotian fauna, consisting of worms and mollusks. Some were only found by using advanced techniques to analyze rocks, which involved laboratory disaggregation of the rocks. It turns out that many of these primitive creatures were missed by previous paleontologists because they were too small to spot with the naked eye.

When we think about it, the first multicelled animals ought to have been microscopic. One-celled life had exclusivity on the earth for about three billion years. Obviously, there was a strong resistance to multicelled life forming for most of the earth's history. Earth scientists believe that having adequate oxygen levels in the atmosphere and oceans was the critical determinant. These multicelled animals used body movement to acquire food and other functions. Movement takes more energy than staying in one place, and oxygen provide the most energy of any reaction. Once conditions were favorable for multicelled life, it would begin with the simplest organisms and develop greater complexity as it evolved into more robust forms. In hindsight, the answer to the Precambrian absence of visible fossils problem was obvious. Look for microscopic fossils of them because complexity comes later. However, the need for that advanced examination was not apparent at first.

Evo Devo

At this point, we shall shift gears and examine a new and important field of research called Evo Devo, which actually helps us understand the Cambrian explosion better. Our current scientific capability for learning about the world is astounding, and Evo Devo is an example of our intellectual power. We learned a lot about life-forms from the Cambrian Period based upon the fossils preserved in limestone shale. Add to that the human ingenuity in extracting information from the fossils to construct three-dimensional replicas of the creatures and then to divine their activities. However, there is more to learn, and the new tool of Evo Devo is the path to that knowledge. Let us see how it works:

One of the frontier areas of the life sciences is Evo Devo, which is short for evolutionary developmental biology. One of the pioneers of Evo Devo and the author of a marvelous book, which explains it in detail, is Sean Carroll. The book is *Endless Forms, Most Beautiful*. Now, Evo Devo is a field of biological research concerned with how different life-forms got to be the way they are and how the processes involved actually accomplish these transformations. For example, the embryos from different animals look very similar to one another, especially in the earliest stages. All of them initially have a head with gill arches. It is almost as if these embryos pass through that ancestral stage of being a fish. Or in another example, consider the stage where an embryo begins to sprout limbs. We have learned that embryos, which will become legged animals, exhibit tiny limb buds early on. Those limb buds will become legs, or will become wings in the case of birds, or will become forelegs and jumping legs in the case of frogs, or will become arms and legs in the case of humans. There appears to be a basic process, which nature uses to construct animals in the image of their parents and that basic process can be tweaked to yield different kinds of body parts.

Another embryonic stage is the development of bodily organs. Embryonic organs appear to develop from three layers known as germ layers. The layers are named the ectoderm, the mesoderm, and the endoderm. These layers are common to all animals, and these layers form the same organs whether bird, reptile, or mammal. One layer forms the heart, another the brain, and so on. No matter how different the adult species may be, the tiny embryos all go through the same stages of development. Again, we sense that a basic process has been modified to produce moderately different results.

We have also learned that hereditary traits are controlled by genes. What we do not fully understand is how genes come into play to build an embryo into a fetus, then a child, then an adolescent, then an adult, and so on. We have learned one very important thing though. It turns out that gene switches within our DNA are vital to the understanding of evolutionary development. Switches are the key to turning genes on and off. It is not until chapter 7 that we explain what DNA is, and how genes are capable of accomplishing tasks for the body. For now, it may help to know that genes can use the machinery in a cell to manufacture chemicals called proteins, and those proteins perform a variety of tasks for the body. Proteins can also turn genes on or off by landing on the gene's switches. Both the genes and their switches reside within the DNA molecules harbored within our cells.

Fruit-Fly Experiments

Scientists have been observing fruit flies for over one hundred years now. Why fruit flies? We are impatient to get results when we conduct genetic experiments, and the life cycle of mammals is simply too long for us. Fruit flies, on the other hand, reproduce very quickly. Moreover, there are ways to accelerate the experiments by increasing the mutation rate of fruit flies. Some mutations involved body parts being duplicated or ending up in the wrong place. For example, legs have appeared where antenna should be or duplicate sets of wings appeared where only one set should be. A great deal of the science of Evo Devo was learned through fruit-fly experiments.

HOX Genes

An important pattern has emerged from many, many experiments on fruit flies and their progeny combined with analysis of the fruit fly's DNA. The fruit-fly larva has a basic body plan, and there is a corresponding zone on their DNA that directly translates to the head of the larva, the middle of the larva, and so on. There are eight genes in a sequence on the fruit-fly DNA that relate to corresponding eight zones on the larva from front to back. Together, this eight-gene sequence comprises a homeobox, and these genes are known as Hox genes.

Once scientists learned of the fruit-fly homeobox, they searched for its counterpart in numerous other animals. What do you think they found? Every animal, which has a body, has a homeobox relationship. Fruit flies have one set of eight hox genes represented in a box. Humans have four sets of Hox genes. Mice have thirty-nine Hox genes. Every animal with a body has Hox genes. Hans Spemann won a Nobel Prize for his work in this area as Hox genes were studied in animal after animal.

Genes from the Distant Past

Our understanding of what we are and how we got here depends on understanding genes and understanding our evolutionary past. Below, we shall consider two important examples of genes and inheritance. Where our eyes came from, and where our ears came from?

Eyes

The ability to see has a life-or-death importance to it in almost every animal species. Vision helps us avoid predators and find food. Except for animals, which live in darkness, vision is universally important to all living beings. The molecule that makes vision possible is opsin. When opsin is activated by light, it starts a neurological chain reaction, which signals the brain. Opsins come in a variety of kinds. One is for black-and-white vision, while others allow us to see in color. Humans have three different color opsins. Birds have four-color opsins and can see into the ultraviolet range. The eye has independently evolved as many as forty times on our planet. As an example of eyes quite different from ours, insects have multifaceted eyes and so did trilobites from hundreds of millions of years ago. Despite there being so many different kinds of eyes, the fact is that all of them utilize opsin for making vision possible.

During the course of some fruit-fly studies, it was discovered that one particular mutation had no eyes at all. And it was found possible to breed progeny flies from it, which also lacked eyes. Could this type of mutation also apply to mammals? Mice are fast breeding so they were tried, and yes, the same results applied in the breeding of mice. Either they were eyeless, had small eyes, or incomplete eyes. Now, we know from medical histories that occasionally human babies are also born without complete eyes. So, having just learned that three species of animals (i.e., fruit flies, mice, and humans) can be born without eyes, it was natural to ask, "Is there a gene common to all three species that controls the making of an eye and whose absence accounts for the eyeless condition?" Enter Walter Gehring, who actually found that gene. The gene has been dubbed Pax 6, and it exists in a similar location on the DNA molecule for all three species. When Pax 6 from a mouse was transplanted into a fly, it grew an eye. Amazingly, it grew not a mouse eye but a fly eye. Gehring also found that he could make an eye grow anywhere on the body; on a leg, on an antenna, or on its back. Other scientists picked up the theme and learned that Pax 6 controls development in everything that has eyes.

Ears

What do you know about your ears? Did you know the ear consists of three components? (1) There is the flap of skin and tissue shaped to direct sound waves inward. (2) There is the middle ear; in the case of mammals, there are three little bones. (3) There is the inner ear; being sensory cells, fluid, and tissue. Incidentally, the external ear flap is a relatively new invention. We mammals are the only ones that have them. Fish, amphibian, and reptiles don't have them.

Another feature unique to mammals is the three-bone combo in our middle ear. German anatomists Karl Reichert (in 1837) and Ernst Guapp (in 1910) traced the two ear bones, malleus and incus, in evolutionary time back to the jawbones of reptiles. This is one of the facts that prove mammals evolved as a branch from early reptiles. Mammal-like reptile fossils are abundantly found in South Africa. The evolutionary stepwise change from one to three ear bones is witnessed in the fossils from that period.

The third bone, the stapes, is traceable back to a bone (i.e., the hyomandibula) in sharks and fish. This bone serves a different function in fish, that of connecting the upper jaw to the skull. The hyomandibula only found application to hearing when certain fish starting spending some of their time on land, where hearing is important. Of course, all that happened hundreds of millions of years ago, and it was a special kind of fish that had evolved lungs. All land animals can trace their ancestry back to these lungfish, and the mechanism for hearing evolved to be more and more useful to survival for them as time went on.

What about our inner ear? It is traceable back to the earliest fishes, where it serves to keep the fish oriented in three dimensions. The organ still serves this function in today's animals in addition to allowing us to hear. Excessive alcohol will disrupt inner ear function causing us to feel like the room is spinning. Some people and animals are born with a defective inner ear, and that is traceable to a defective gene called Pax 2. Studies show that Pax 2 is common in today's animals and those in the evolutionary chain.

How We Can Assign Genes to the Long-Extinct Cambrian Animals

At first blush, it seems impossible, doesn't it? How in the world could we ever know what genes existed in the DNA of those first animals to leave impressions of their bodies in the limestone shale of million years ago. And yet, Sean Carroll and his team did that very thing. First of all, he united what he knew of the fauna of the Burgess shale with that of the Chengjiang fauna. This latter group are the same marine animals but from a time fifteen million years earlier than the Burgess shale fauna. The thing that sprang to mind for Carroll was that the combined fauna fell into two distinct groups, arthropods and lobopodians. Perhaps the most distinguishing feature separating them is that arthropods have jointed legs and lobopodians have unjointed legs. The following table lists some of the Cambrian fauna, which fall into these two groups.

Table 2.1 Arthropods and Lobopodians of the Cambrian Fauna

Arthropods	Lobopodians
Marella	Aysheaia
Olenoides	Mictrodicton
Waptia	Opabinia
Yohoia	Anomalocaris

Is it possible that one of these animals still lives today? Carroll selected a type of animal that comes close to matching the Burgess fossil Aysheaia and still walks the earth. Its name is Onychophora, and it is a lobopodian. His team had to go to Australia to find them, but they were successful in their quest. Next came analysis of the Onychophora genome. They focused on finding the Hox genes and were pleased to learn that Onychophora had all the Hox genes found in modern-day fruit flies and arthropods. Moreover, because arthropods are believed to have evolved from lobopodians, we can conclude that the Cambrian fauna had these same ten Hox genes too.

I hope my readers appreciate what a huge discovery this was for science. Everything that we have learned in assembling the knowledge and tools of Evo Devo from studying living animals is now valid for all animals back to the beginning of the first animals. Not too shabby, eh?

Three

Life Explodes in the Phanerozoic Eon

The Chapter in a Nutshell

Once multicelled life burst forth in the Cambrian period, the fossil record, of the intervening 560 million years between then and now, presents us with rich and complex evidence of continuously evolving plants and animals. Geologists have named this time span the "Phanerozoic Eon," and it is the focus of this chapter. Darwin had shown us how natural selection acts on species to optimize their form and behavior so they can survive and flourish in a competitive landscape. Yet, the actual mechanism of trait inheritance eluded him. How did the fittest pass their superior traits to their offspring? Other researchers stepped in and by the mid-twentieth century had solved the mystery and unleashed the power of genetic science. Genes are the key to the puzzle of inheritance and also the key to the puzzle of evolutionary change. Species can evolve to adapt to their changing environments because favorable mutations in their genetic code lock in those incremental changes. This process accounts for the progression of different life-forms, which we see in ancient fossils.

Geologists have established from rock fossils that extinctions have happened throughout geological history. Of particular interest are five major global mass extinctions since the Cambrian period. Existing life-forms tend to be wiped out on a wholesale basis by these major extinctions, but after a pause, new plants and animals evolve to fill the empty ecological niches. These major transitions define distinct geological periods uniquely identified by their fossils. The most severe global mass extinctions separate the geological eras. Thus, geologists have assigned three different era designations to the time since fossils were first visible. They are called the Paleozoic, Mesozoic, and Cenozoic eras. The Paleozoic is the era, where arthropods appear in the first period (Cambrian), fish evolved to fill the oceans, amphibians evolved from lungfish adapting to land, and reptiles evolving from amphibians to break their dependence on being near water. The End-Permian extinction decimated much of the flora and fauna to bring this Paleozoic era to a close. The Mesozoic era followed with its dinosaur dominance. Pterosaurs and sea reptiles were kings of the sky and oceans, respectively. The K/T Extinction brought an end to the long reign of these animals and brought a close to the Mesozoic era. The new era, called the Cenozoic era, is the time we live in. The empty ecological niches resulting from the global extinction were mainly occupied by two kinds of warm-blooded animals; birds and mammals.

The Ancient Past Lies at Your Feet

In this chapter, we shall look back into the time before people ever existed on this planet. We already discussed the strange animals that appeared in mass during the Cambrian Period. Plant and animal life had just gotten started. The evolutionary process would build animals at various stages over millions of years, which swam in the seas, crawled onto land, became independent of water for reproduction, dominated the land, and step by step certain of them became more like us. Many of the genes invented by nature for these distant ancestors reside in our human bodies still today. Reminders of these ancestors exist among us, but we may not even notice. Here are some examples.

Coal

West Virginia, once a strongly Democratic state, became a Republican state when the coal industry was eschewed by the clean-energy crowd. Although, automation is the miner's biggest nemesis. Whatever the cause, coal miners were out of work and hurting economically. The Appalachian Mountain country reaches beyond West Virginia, of course, and so the coal-miner problem extended into Pennsylvania and Kentucky. This story relates to our distant past. The Carboniferous Period (360–290 mya) of the Paleozoic era is with us in this modern story, for it was back then, when all these compressed plants, called "Coal" were formed. We know that coal is compressed plant life because occasionally the details of ancient leaves are preserved.

Dinosaurs

Have you ever been to the Grand Canyon in Arizona? Everyone is impressed by the majestic views of these multicolored strata. However, fewer viewers realize that the ground they walk upon is over 250 million years old. That is right, in a way you have returned to the later days of the Paleozoic era. In fact, if you search diligently, you may find marine fossils from that time period. Erosion has exposed these ancient strata by removing the overload that had been laid down later. Less erosion happened in the Petrified Forest farther east in the state. For those fallen and fossilized trees were alive in the Triassic Period of the Mesozoic era, the so-called age of reptiles. The first dinosaurs roamed the land in the Triassic, and nature has preserved their footprints in Southern Utah. If you want to get to more recent times in the Mesozoic, drive north. The Escalante staircase is in one sense a journey through the Mesozoic era, thanks to varied degrees of erosion. The Jurassic Period is the middle period, and the Cretaceous Period was the final period of the Mesozoic. Utah and Montana have been a large source for dinosaur fossils from these times.

Mary Anning

Jumping from the American West across the Atlantic Ocean to the British Isles, we come to the story of Mary Anning. "Who?" you ask. Well, perhaps you have heard the little tongue twister that describes her—namely, "She sells seashells by the seashore." What she sold was more valuable than seashells though. Mary was born in 1799 in Lyme Regis, England. Her father was a cabinetmaker, but he was also an amateur fossil collector. Unfortunately, he died when Mary was only eleven and left Mary and her family with heavy debts and no income. Young Mary's resourcefulness at this juncture makes me think of the saying, "When the going gets tough, the tough get going." Young Mary actually raised money to help support her family by finding fossils and selling them. She and her brother found the first ichthyosaur (a porpoise-like marine reptile) and the very first pterodactyl (a flying reptile) among many other important finds. These creatures lived along with the dinosaurs in the Mesozoic era.

Figure 3.1 Ichthyosaurs, Marine Reptiles of the Mesozoic Era
By Heinrich Harder (1858–1935)—The Wonderful Paleo Art of Heinrich Harder, Public Domain, https://commons.wikimedia.org/w/index.php?curid=838851.

A wealthy collector, Thomas Birch, befriended Mary's family and even sold off parts of his collection to assist them. Over time, Mary learned the science and the art of fossil restoration. In fact, Mary could hold her own with the best experts in the field. Moreover, she was skilled at finding impressive fossil animals. So much so that Mary Anning became famous throughout Europe due to her impressive finds. The British Association for the Advancement of Science awarded her an annual stipend in appreciation of her contributions to science. In 1847 Mary succumbed to breast cancer.

For those of you curious about ichthyosaurs, you might be interested in learning that these creatures predate the dinosaurs by ten million years. Although they resemble whales and dolphins in appearance, they were reptiles, not mammals, and their tails moved sideways like fishes. Whales and dolphins move their tails up and down. There were many species of them, but huge eyes and a long, narrow snout are typical trademark features.

The Grand Karoo

Then there is the Grand Karoo, a fossil-rich area in South Africa. It has been a fossil gold mine for paleontologists specializing in the origins of mammals, tortoises, and dinosaurs. It is rich in fossils from the Permian Period (the last period of the Paleozoic era) and the Triassic Period (the first period of the Mesozoic era). The origin of mammals is found here in Therapsids, a mammal-like reptile. There is also fossil evidence of the first dinosaurs in the Karoo. This area also contains evidence of two global mass extinctions: the End-Permian extinction (252 mya) and the End-Triassic extinction (200 mya).

All Life Is Related

There are at least three reasons to believe that all living things today are related to each other. First is the fact that all living things can trace their ancestry back to LUCA, the last universal common ancestor. Second, we also know that all living things are based on DNA, which is a unique and complex molecule. It seems highly doubtful that life containing such a complex biochemical substance could have originated more than once. Moreover, we have the technology today to compare the genomes of any two species and based on their genetic distance, get a good estimate of when their common ancestor existed. The third argument for relatedness is that multicelled plants and animals require an energy system based upon oxidation. We do not see them in the fossil record before 800 million years ago, because there was not enough oxygen in the environment until then.

When we consider all the different types of plants and animals in the world today, it is a huge number. Just enumerating all the different insects in the world would be a Herculean task. However, considering the Earth is 4.5 billion years old, these creatures are relative newcomers. Multicelled life could not get started for close to three billion years even though single-celled life arose as soon as the planet cooled sufficiently to allow it. However, once that it got started, it radiated into countless forms. The process is called speciation, and it is one where an existing species gives rise to two or more new species over time. Those new species give rise to even more new species, and before long, the world is filled with millions of species.

Competition and the battle for survival is a brutal business, and it is true that most species eventually go extinct. Global mass extinctions accelerate the culling process and may wipe out a large percentage of the existing species before it is done. Extinctions have knocked the number of species back to a much smaller number at least five times since multicellular life arose on the planet. Happily, life always bounces back. At least, it always has so far. New kinds of plants and animals arise during each new recovery. Through all this chaos, one thing remains true. All of us can trace our ancestry back to LUCA, and all of us are related through a common ancestor to each other.

Evolution and Genetics

In the previous chapter, we introduced Charles Darwin and his theory of evolution. Please don't let the word "theory" mislead you. Skeptics love to say it is only a theory, implying that it isn't necessarily true. If the word were "hypothesis" instead of "theory," their argument might be valid. However, a scientific theory has a much higher status than a hypothesis. A theory has weathered skeptical scrutiny for decades and has still proven to be sound. The theory of evolution, which has endured for over a century, is essentially a law. In fact, it is the very foundation of the biological sciences. With that said, we pick up the thread of Darwinian evolution here and introduce the concept of genes. No doubt, you already have some inkling of what a gene is or for all I know you may be an expert on the topic. I am going to assume the average reader isn't an expert and proceed from the basics. Genes are vitally important to understanding evolution, and evolution is the topic of this chapter and of the book.

Now I grant you, that Darwin's basic theory has been modernized over the years as biological discoveries have been made that were applicable. In particular, the science of inheritable traits (i.e., genetics) was not understood in Darwin's days. Back then no one understood the mechanism by which parents pass their traits to their children. Obviously, sex is a vehicle for transporting those traits, but what is the biological mechanism? Darwin, himself, wasn't sure how traits were biologically transmitted to one's offspring. It was obvious that both the mother's and the father's traits were visible in offspring, but were their contributions blended? As Darwin pondered this question, the real answer was being developed on the European continent. A German monk named Gregor Mendel was discovering the biological laws of inheritance through plant-hybrid experiments, but Darwin never connected with him, and Mendel's concepts faded into obscurity for a time.

It was not until the early twentieth century that Mendelian genetics were rediscovered and took center stage. Mendel had postulated a basic unit of inheritance that he called the gene. The gene could control a single trait such as whether you have blue or brown eyes. The concept of the gene provided the vital element that made natural selection a valid theory. You might want to apply highlighter to that last sentence! It meant that a trait that has strong survival value can be passed to one's offspring undiluted. The gene that governs the trait is discrete and indivisible.

The science of genetics made discovery after discovery in the early twentieth century. Genetics advanced rapidly as better microscopes and experimental techniques were invented. Moreover, more knowledgeable and better-educated scientists were working on these problems and sharing information. By the middle of the century, scientists were poised to make the greatest breakthrough ever. They were ready to discover the structure of DNA, the mystical material in our cell's nucleus that carried the stuff of inheritance. Within the immensely long DNA molecule live our twenty-thousand-plus genes. By the 1960s, DNA was recognized as the regulator of life and its double-stranded helix structure established. Today, DNA science is providing mind-boggling power to humanity. In fact, it has provided godlike power to humans. We will touch on this later in the book.

Genes

What Is a Gene?

As much as evolution is a thread running through this book, the concept of genes is another thread. These two threads are closely linked. Whereas evolution is a theoretical concept that explains the parade of plants and animals through geological time, genes are the mechanism by which these evolutionary transformations occur. Genes control inheritable traits as Mendel first suggested. So, in a collective sense, genes explain the differences between species. The offspring of cats are kittens, the offspring of sparrows are baby sparrows, and so on. Genes are responsible for this. Now, the creatures we encounter, accomplish this reproduction via sexual unions. The genes of the father are combined with the genes of the mother in a process, we call conception. We can mentally visualize individual sperms swimming toward the mother's egg in a race to be the lucky one to fertilize the egg. The lucky sperm that succeeds in fertilizing the egg, carries the father's genes. The egg carries the mother's genes, and when they unite, the embryo will have a unique mixture of its parents' genes. We humans have over twenty thousand genes in our genomes, so the process of assembling the embryo of a new human being is complex. Of course, it is equally complex for other animals too.

What is a gene exactly, and how does it work? I am going to answer that question simplistically for now. We will dig a little deeper into the definition of a gene later in the book. So, in brief, genes are made of the material you know of as DNA. That material carries a four-letter code that the cells of your body can read and act upon. Genes are instruction codes provided to your cells for synthesizing a host of different proteins, which in turn build body parts, regulate bodies, and the like. We now have an idea of how living things function. The life-force is essentially a biochemical phenomenon, which is at once ingenious, incredibly complex, and wondrous. How lucky we are to live in an age where these processes have been revealed to us. We can thank the tireless efforts of researchers, some still living, but many long gone.

What Drives Evolution?

Evolutionary change in a species is a gradual process, which occurs over numerous generations. Some sustained, long-term stressor has to drive the change, otherwise a species stays the same. For example, the climate may change, usual foods may disappear, new predators may materialize, and so on. The stressor is severe enough to present a life-or-death struggle for at least some of the population. This results in a culling of the less adaptable but spares the better-adapted individuals. It is the diversity of the population, which determines the severity of the culling process. However, for a species to really be able to live for generations under a continuous and relentless stress, permanent changes in their genomes are needed. They need some new genes, which fundamentally change their ability to cope with or overcome the stressor. Here is how that happens: That beautiful biochemical system governing inheritance, which we described above, sometimes makes mistakes (called mutations). Most mutations are not helpful, but when the rare useful ones occur, those new beneficial genes tend to gain dominance in the generations that follow their first appearance. Concurrently, the less beneficial genes tend to disappear over generations. This in essence is how evolution works in changing species. It can be viewed as a battle between genes taking place over many generations.

Another point I want to make about genes is that they exert their influence upon our world more powerfully than you might imagine. Richard Dawkins is a prolific science writer, who has argued this point more effectively than anyone else. His landmark book *The Selfish Gene* changed the way scientists thought about evolution. From a species viewpoint, it is not the individuals of the species that are important, it is their genes. Dawkins argues that genes are close to being eternal entities, whereas we individuals are merely vessels by which our genes pass themselves forward into future generations. It is the competition among genes that is the significant contest. Much of what we see in nature is due to that competition.

After multicelled life arose on our planet, over one-half billion years of genetic experimentation in nature has been taking place with lifeforms adapting to the ever-changing environment. The very fact that new plants and animals arise amounts to a change in the environment because new competitors have joined the game. It may seem impossible for a four-legged animal to transform into a whale, for a fish to transform into a frog, or a reptile to transform into a bird, and yet the fossil record tells us that these transformations actually occurred. Our own genomes contain genes from ancient fishes, amphibians, reptiles, and mammals. You see, Dawkins was correct about genes being nearly eternal. This chapter will discuss some of those unbelievable transformations.

Stratigraphy

This chapter is about the Phanerozoic Eon, which began with the Cambrian period and contains within it the Paleozoic era, the Mesozoic era, and the Cenozoic era. We know that this eon is full of extinctions and adaptive radiations, where species come and species go. We just discussed how genes make those specie's adaptations possible. However, we are not ready to tell that story of the Phanerozoic Eon yet. You see, the story is also one of geology. Those plants and animals kept changing due to a planet's surface that kept changing. Things are no different today. Hurricanes, tornados, wildfires, floods, tsunamis, earthquakes, and other disasters remind us that the planet has an agenda not in keeping with our own. So next, we look at the world as a geologist sees it.

Superposition

How do we know what the Earth was like in the distant past? Actually, scientists have a variety of ways of investigating the history of the Earth. One of the most informative ways is through the examination and dating of fossil plants and animals. As you know from viewing landscapes like the Grand Canyon, Bryce Canyon, and others, the Earth is composed of different strata, lying one on top of the other. Closer examination of these strata may reveal fossils that uniquely identify them. We have also established that a lower stratum is older than the one above it and younger than the one below it. This principle is called the Superposition principle. In principle, any particular stratum ought to be sandwiched between its upper stratum and its lower stratum as we search for it around the globe. However, processes like erosion remove strata differently at different places on the planet, so you may not be able to find it. The superposition principle combined with the uniqueness of strata based on their fossils can provide us with a chronological account of the earth's history. The story we want to tell of the Phanerozoic Eon is recorded in the strata.

Speciation and Extinction

Now, it is a fact that living things have changed in form over geological time. The vast majority of species, which once existed, have gone extinct. Paleontologists have carefully correlated the various species to the strata in which they have been found and thereby determined their first appearance and eventual disappearance from our fossil records. We also see transformations of the species in the fossil record. For example, horses, asses, and zebras can be shown to have descended from a more primitive ancestor having three toes. The strata record these changes in form, and scientists have been able to establish a chronology for life going back to the first visible fossils. Geologists have studied the sedimentary rock layers from around the world and named the individual layers uniquely identified by the fossils found in them. Using the superposition principle, they have created a historical record of the Earth.

Geology as an Emerging Science

If you were a person living in the eighteenth century, it is likely you either believed that the unusual animals displayed in the ancient strata was the result of the Biblical great flood, or you joined the camp that believed the strata represented a deep history of changing animal forms. James Hutton represented the latter school of thought. The scientific revolution was underway in many fields, and Hutton was a leader in establishing geology as a scientific field. For example, he thought about the Earth's hot interior and postulated that it was the source of new rock. He expounded on his interpretations in a paper delivered to the Royal Society of Edinburgh in 1785. Hutton's theories pointed the way to new discoveries and that motivated many Earth scientists to go investigating for themselves. By the early nineteenth century, the scientific community was generating a wealth of information concerning strata and the unique kinds of fossils found in them. This became the criteria for identifying a particular stratum anywhere on the globe. If two strata contained the same fossils, then those strata were from the same time in history when those animals lived.

Worldwide Correlation

British geologists dominated this correlation and identification process and were responsible for naming the different periods. They also noticed that certain sequences of strata were repeated in different regions of the world. A collaborative process, led by Abraham Werner, was undertaken to correlate this global information. The result was a chronology of life-forms found throughout the world. They divided the strata into four chronological sections, namely, primary, secondary, tertiary, and quaternary. We have a different naming system today, but the old one is often referred to in the literature.

Table 3.1 Geological Time Chart

EON	ERA	PERIOD	MILLIONS OF YEARS AGO
Phanerozoic	Cenozoic	Quaternary	1.6
		Tertiary	66
	Mesozoic	Cretaceous	138
		Jurassic	205
		Triassic	240
	Paleozoic	Permian	290
		Pennsylvanian	330
		Mississippian	360
		Devonian	410
		Silurian	435
		Ordovician	500
		Cambrian	570
Proterozoic	Late Proterozoic Middle Proterozoic Early Proterozoic		2500
Archean	Late Archean Middle Archean Early Archean		3800?
Pre-Archean			

https://upload.wikimedia.org/wikipedia/commons/2/2c/Geologic_time_scale.gifWikipedia commons.
Thanks to the United States Geological Survey (http://pubs.usgs.gov/gip/fossils/fig15.gif) [Public domain], via Wikimedia Commons.

Plate Tectonics and Continental Drift

"Plate tectonics" is a theory about how the outer layers of the Earth (i.e., the lithosphere) are composed of about a dozen plates, which move somewhat like slabs of ice on a lake. Of course, the landmasses are not ice slabs, and the underlying material is not water. Instead, the landmasses (i.e., plates) are rigid rocklike structure, and the underlying material is the hot, plastic material of the Earth's interior (i.e., the asthenosphere). Convection currents exist in the asthenosphere capable of moving the continental plates, albeit very slowly. In my youth, this theory was deemed ridiculous. How could solid rock continents of immense weight be made to move? Today, there is overwhelming evidence that it is true, and the vast majority of scientists agree with the theory.

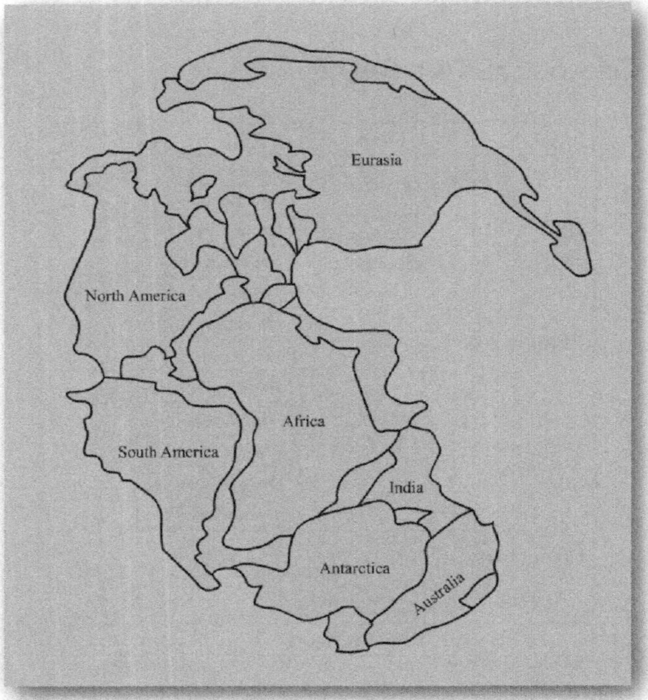

Figure 3.2 Pangaea, Supercontinent
CC BY-SA 3.0, https://commons.wikimedia.org/w/index.php?curid=48962.

"Continental Drift" is a theory that explains how continents can move on the surface of the Earth. It was first proposed by Alfred Wegener in 1912 but was not taken seriously until decades later. One consequence of continental drift theory and much evidence is the concept of a supercontinent, which formed about 300 million years ago. This supercontinent was called "Pangaea" and became important to us in understanding what forced influenced events of these distant times. The supercontinent started breaking up into the continents, which we see today, about 100 million years ago.

Mass Extinctions and Adaptive Radiation:

What was it that the geologists observed in the strata that made them want to divide the past up into distinct eras, periods, and epochs? It was dramatic differences in the flora and fauna fossils between one stratum and the next. They became aware that old established plants and animals seemed to disappear and that new kinds of plants and animals replaced them. The major divisions of Paleozoic, Mesozoic, and Cenozoic eras were the most dramatic in their differences, and the cause was global mass extinctions. The most severe of these global mass extinctions, the End-Permian Extinction, ended the Paleozoic era and initiated the Mesozoic era.

Evidence of Mass Extinction

Imagine yourself as an amateur geologist exploring a nearly vertical rock wall with sedimentary strata exposed. There you are in sturdy boots, hiking clothes, backpack, with your rock hammer and magnifying lens. You are interested in a white upper stratum, a black ribbon below it, and a sand-colored stratum below the ribbon. You use the hammer to chip out rock fragments and search for fossils in them. You record what you see in your field notebook. After several hours of such work, you sit on a large nearby rock, take out a sandwich from your backpack, and chew your food as you review your notes. Your overall observation is that very different kinds of animals lived in the white stratum than lived in the sandy stratum. The black ribbon had no fossils at all. You have just seen physical proof of an extinction event. Some animals survived the extinction and replenished the area with descendant species. If you can duplicate these findings at several sites around the world, you are probably well on your way to establishing a global mass extinction. Extinctions come in a variety of sizes, but there have been five global mass extinctions since fossil animals appear in the fossil record. We discuss them next.

The Big Five

There are five mass extinctions, which rank as the worst extinctions of all time. Table 3.2 summarizes their effects:

Table 3.2 The Big Five Extinction Events

Extinction Event	When (years ago)	Total families before event	Families lost in event,
Late Ordovician	440,000,000	450	22 %
Late Devonian	360,000,000	450	22%
Late Permian, a.k.a. P/T	250,000,000	400	50%
Late Triassic, a.k.a. T/J	213,000,000	300	20%
Late Cretaceous, a.k.a. K/T	66,000,000	650	15%

Note: The taxonomy order is kingdom, phylum, class, order, family, genus, and species.

These major extinctions are presented in chronological order.

Late Ordovician

The vast majority of animal life lived in the sea, and this extinction took a heavy toll. Eighty-six percent of life was wiped out. The cause of the extinction was glaciation, and some call this period "Snowball Earth." Sea levels fell as did oxygen levels (anoxia). Trilobites, brachiopods, and graptolites were hard hit.

Late Devonian

As life came back in the Devonian period, it was the same Ordovician creatures. This is not what typically happens. Two supercontinents, Gondwana and Euramerica, occupied the lower hemisphere, and the rest of the planet was ocean. The Devonian ended with a major extinction, which wiped out 75 percent of species. This was the second-worst mass extinction in the Earth's geological history. Like the previous mass extinction, this one too was due to anoxia in the seas. Trilobites barely escaped being totally eliminated. Large land plants had evolved during the Devonian, and it is thought that the nutrients from their roots had washed to sea and caused algal blooms to form. This led to the anoxia of the seas.

Late Permian

The Late Permian extinction is considered to be the mother of all mass extinctions because it wiped out from 76–96 percent of all species and 50 percent of all families. Volcanic activity was a major cause of the extinction. Massive amounts of carbon dioxide were released into the atmosphere, leading to global warming, acidic oceans, and anoxia. Anaerobic bacteria created what are known as Canfield seas, nearly depleted of oxygen. Some of these bacteria process sulfur instead of oxygen and release toxic hydrogen sulfide as a waste product.

The Permian-Triassic extinction caused severe reduction in diversity, especially among marine invertebrates. Sampling of Chinese strata indicates that 286 out of 329 marine invertebrate genera were lost in the extinction. On land, amphibians and reptiles were hard hit. Moreover, this is the only extinction to ever severely affect terrestrial insects. Mammals had not yet evolved in the Paleozoic, but the fate of future mammals was at risk due to this extinction. Mammals are descended from a mammal-like reptile or therapsid. The therapsids were especially hard hit by the extinction because they had evolved under higher oxygen times, and they relied on an ample oxygen supply to exist. One genus of therapsid not only survived but flourished, namely the lystrosaurus. It had evolved a huge barrel chest with increased lung capacity and a superior blood circulation system. This extinction ended the Paleozoic era and ushered in the Mesozoic era.

Late Triassic

Marine life came back after the devastating End-Permian extinction, changed and more complex. Snails, urchins, and crabs appeared as new species. The extinction, which ended the Triassic period, may have been due to both a meteor impact and heavy volcanism. It was milder than the End-Permian by far. Twenty percent of families were wiped out compared to fifty percent.

Late Cretaceous

The Cretaceous period was the last period of the Mesozoic era, the age when dinosaurs dominated the landmasses. They weren't the only dominant reptiles though; pterosaurs dominated the skies, and marine reptiles were mighty predators of the seas. Our mammalian ancestors coexisted but were squirrel-size and mostly nocturnal. This reptilian dominance had endured for 185 million years but came to a brutal end with the End-Cretaceous extinction. Now, this extinction was the focus of a major controversy as scientists took sides on its cause with either the meteor impact crowd or the volcanism crowd. Overwhelming proof that an impact had occurred was presented. The element iridium was found in the fossil-less stratum separating the Mesozoic and Cenozoic eras in abnormal amounts. Meteorites are high in iridium. The impact crater was even identified by tracing the geological clues to their logical conclusion. That crater is located in the Yucatan peninsula of Mexico. The K/T stratum also contained noticeable carbon. Apparently, red-hot ejecta from the impact descended on the forests of the world and set them ablaze. It is also thought that the finer dust particles stayed in the upper atmosphere for weeks, blocking the sun and halting photosynthesis. Since the time of that victory for the impact crowd, the debate fever has cooled down, and it is now agreed that the heavy volcanism occurring in the Deccan traps of India had contributed to the death toll too. This K/T extinction is especially meaningful to us because the impact wiped out the dinosaurs and allowed our mammal ancestors to fill the empty niches.

Hox Genes and the Emergence of Vertebrates

In chapter 2, we discussed the power of the new scientific field, called Evo Devo. It unites the science of embryology with the science of DNA and genetics to give us a new understanding of the evolution of living things and an understanding of what we observe in the fossil record. Evo Devo helped us understand how so many diverse species sprang into being during the Cambrian explosion. Arthropods, such as trilobites, were the dominant type of animal in the Burgess shale, and its diverse forms are explained by the ten Hox genes that govern how different zones of the animal will develop. For the early animals of the Cambrian Period, that cluster of ten Hox genes was sufficient and still is today in the insects, spiders, crustaceans, and the like that are descendants. The problem is that, for animals, like ourselves possessing a skeleton, ten Hox genes are far too few.

The problem was solved early in the Paleozoic era. Remember that fish are vertebrate animals, and they appear shortly after the Cambrian Period. Scientists see strong evidence that the set of Hox genes doubled, and then a bit later, doubled again. The net effect was a fourfold increase in the number of Hox genes. When duplicate genes exist, they eventually get put to a new use. For vertebrates, that meant a way to build cartilage and bony structures. It also meant a way to build superior and more complex brains. When scientists examined the genomes of mammals, birds, and many types of fish, they found four Hox gene clusters. So, we can conclude that the four clusters existed in the common ancestor of these vertebrates, whereas all invertebrates have but a single cluster.

The Periods of the Paleozoic Era

The Cambrian

The Cambrian period began 541 million years ago and ended 485 million years ago. This period is noted for a burst of numerous of new creatures on the scene seemingly out of nowhere. It is such a remarkable event that it carries the name, the Cambrian explosion. This period and its animal life were discussed in chapter 2. A point of geological interest is that the supercontinent Gondwana had been forming during the Cambrian period. It was situated from the equator to the South Pole. The warm climate and moist conditions of the Cambrian were ideal for the radiation of species.

*Note: Gondwana was composed of what we know as Africa, South America, Australia, Antarctica, and Siberia. Laurentia was composed of North America, Baltica was composed of northern Europe, and Avalonia was composed of Western Europe.

The Ordovician

The Ordovician period began 485 million years ago and ended 443 million years ago. This was the period in which primitive fish, cephalopods, and corals evolved. Cephalopods are marine creatures. Those living today include squid, octopuses, cuttlefish, and the nautilus. There are about eight hundred species of cephalopods living today. The most common marine animals in the Ordovician were trilobites, snails, and shellfish. The first land animals evolved in this period too. These were arthropods (i.e., insects, crustaceans, and others). Arthropods also dominated the seas. As for the planet, you might be surprised to learn that a map of the Earth back then is nothing like what we have today. You would see continents named Laurentia, Baltica, and Avalonia. And there was also the southward-drifting supercontinent Gondwana, which had an ice cap on it, and it would have looked a lot like Antarctica does today. The Ordovician period was a time of extreme climates, from greenhouse conditions to icehouse conditions. The period ended with a horrific mass extinction. Glacial conditions became so severe many of the newly formed species went extinct. Some refer to this phenomenon as "Snowball Earth."

The Silurian

The Silurian period began 443 million years ago and ended 419 million years ago. Life began to return after the devastation of the Snowball Earth event. The jawless fish of the Ordovician period proliferated again, but the first jawed fish also appeared during this time. During the Ordovician, fish could not survive in freshwater, but new species that could survive began to appear during the Silurian. It is also noteworthy that animal life on land got a much better hold than in the Ordovician, although they were arachnids and centipedes. Arachnids are eight-legged creatures such as spiders, scorpions, ticks, and mites. When I imagine the land crawling with these creatures, I don't think that I want to take a time machine back to the Silurian anytime soon. Another significant development of the Silurian is the development of lignin tissue in plants. It was significant because of the greater structural properties it contributed, allowing plants to grow larger and for them to make huge inroads on land for the first time.

The Devonian

The Devonian period began 419 million years ago and ended 359 million years ago. When you think of the Devonian Period, think of fishes. A huge diversification of fishes occurred during these times, and one of those new kinds of fish was our ancestor. Scientists believe it was the lobe-finned fish, which led to the evolution of the tetrapod, the first land vertebrate. These amphibians had an adaptive radiation of their own. The vascular plants, which had evolved in the Silurian, had such a tremendous burst of adaptive radiation, that the event is known as the "Devonian Explosion." Carbon dioxide levels fell, and plants compensated by evolving leaves capable of drawing in more of it. The period ended in a mass extinction event called "the Late Devonian Extinction," in which 70 percent of all species went extinct.

The Carboniferous

The Carboniferous period began 359 million years ago and ended 299 million years ago. The climate was quite warm in the early Carboniferous period, and tropical swamps dominated the landscape. There was no better time in the history of the Earth for plant growth. Oxygen levels were 80 percent higher than today. The bacteria and fungi, which hasten the decomposition of dead plants today, had not yet evolved, and much of the abundant vegetation was buried, hence the huge coal reserves available to us today. The global map was still changing—a collision of the continent Laurussia (Europe and North America) with the continent Gondwana producing the Appalachian Mountains.

A significant event in the animal chain, which led to us, was the development of the amniotic egg. While it was true that amphibians had well adapted to a life on land, the fact that reproduction of their species had to occur in water meant they had to reside close to bodies of water. The invention of the amniotic egg eliminated that restriction. Reproduction and raising of the infants could now occur anywhere on land. All descendants of these innovating animals are called amniotes, and that includes you and me.

The amniotic egg led to the evolution of the first reptiles and synapsids. What is a synapsid? It is essentially those amniotes, which are not reptiles. Defined a bit differently, it is those primitive amniotes, which eventually became mammals. However, a paleontologist would define the synapsids as those animals with a single small opening in the skull behind each eye. The opening is called a temporal fenestra. This name makes some sense to me because I remember from German Language class that "fenster" meant window.

The Permian

The Permian period began 299 million years ago and ended 252 million years ago. The land animals alive in the Permian were a diversity of arthropods, amphibians, and reptiles. Many of those reptiles were the cold-blooded, small-brained synapsids of the previous period. Toward the end of the Permian, the first archosaurs evolved. They would become the dinosaurs of the Triassic Period to follow. This period was the last period of the Paleozoic era It was the most severe global mass extinction in geological history. The fact that the world's continents had combined to form the supercontinent Pangaea may have been a factor in the extinction.

(Source: https://en.wikipedia.org/wiki/Paleozoic).

The Mesozoic Era

We often refer to people and things, which are obsolete, as "dinosaurs." They are out of touch with today's world and no longer fit in. Yet, a representative for the real dinosaurs might argue that mammals have only been dominant for 65 million years, and humans have only been dominant for a tiny fraction of that span. Dinosaurs, on the other hand, were dominant for around 180 million years. So, as we discuss the Mesozoic era, we will focus on dinosaurs, who were the dominant land animal. However, there were also amazing animals in the air and in the oceans. They are next.

The Pterosaurs

The pterosaurs were flying reptiles, sometimes reaching the size of a small airplane. They first arose in the late Triassic and continued evolving in the Jurassic and Cretaceous. They all went extinct in the End-Cretaceous extinction. The earliest pterosaurs had long tails, jaws with teeth, and their wings were membranes of skin and other tissue stretching from ankle to the end of a highly extended finger. Pterosaurs are presumed to be warm blooded, and the fact that they had furry coats tends to confirm it. Later pterosaurs evolved shorter tails and fewer to no teeth.

Figure 3.3 Ramphorhynchus, a Jurassic Pterosaur
By Charles Whymper (1853–1941) (Public domain), via Wikimedia Commons

Marine Reptiles

During the Mesozoic era, dinosaurs ruled the land, pterosaurs ruled the air, and some monstrous, air-breathing, marine reptiles ruled the seas. They ranged in size from three feet to fifty-six feet in length. They are known as ichthyosaurs, plesiosaurs, pliosaurs, mosasaurs, and others. Ichthyosaurs evolved in the Triassic from land reptiles that returned to the sea, and they finally went extinct in the late Cretaceous period. Plesiosaurs are recognized by their long necks, small heads, sleek bodies, and broad flippers. They were the sea serpents of myth except they were all extinct long before humans evolved to witness them. Pliosaurs resemble plesiosaurs except they had short necks. Mosasaurs evolved in the Cretaceous and gave live birth to their young in water. Indications are that Mosasaurs were warm blooded. Their impressive tails did the work of propelling them through the seas.

Figure 3.4 A Swimming Plesiosaur
By Shiqiu Liu [CC BY 2.5 (http://creativecommons.org/licenses/by/2.5)], via Wikimedia Commons.

Some Interesting Features of Dinosaurs

Better Breathing

The most distinctive feature of dinosaurs is that their legs are oriented directly under their torsos. This new skeletal design was a huge improvement over the design of the reptiles, which preceded them. Consider how today's lizards get around to help you visualize this: Lizards have their legs sticking out nearly horizontally from their bodies. In order to walk or run, lizards must twist their spines through an "S" pattern. When its torso twists, the compressed side cannot suck air into that lung, and when the opposite side is compressed, the other lung is similarly restrained. So for half the time, the lizard is only getting the benefit of half of its total lung capacity. Dinosaurs, by contrast, do not have to contort their spines in order to move because their legs are situated directly under their bodies. They can use 100 percent of their lung capacity. Full lung function is vitally important for chasing down prey or even worse, running from a predator.

Figure 3.5 Gasosaurus Constructs
By Paleocolour (Own work) [CC BY-SA 4.0 (http://creativecommons.org/licenses/by-sa/4.0)], via Wikimedia Commons.

Note: Gasosaurus was a carnivorous dinosaur from the mid-Jurassic period around 164 million years ago. It is estimated to have been eleven to thirteen feet in length and to have weighed between 330 and 880 pounds.

Bipedal Locomotion

Here is another fact about dinosaurs: most of the land animals, we see at the zoo, walk on four legs. They are quadrupeds. However, the earliest dinosaurs walked on their two hind legs. They were bipedal. Their hind legs were much bigger and stronger than their forelimbs. Now, unlike us, dinosaurs had a body oriented horizontally. So, in order to maintain their balance, they needed to have weight in back of their supporting legs equal to the weight in front. The solution was to have a long, thick, and very stiff tail. One design improvement often leads to another improvement, and that was true for the dinosaurs. Now that the hind legs were directly below the torso supporting the animal's weight, there was no reason that its legs couldn't be longer. The benefit of longer legs is longer gait and that made it possible to run much faster. Many new species of dinosaurs radiated from this origin body plan. A few of them evolved into quadrupeds. Especially dinosaurs, which carried armor plating, needed all four legs to support the additional weight. Even so, these quadrupeds had short forelimbs and long hind limbs.

Cold Blooded or Warm Blooded?

Were dinosaurs warm blooded? They were probably about half-way between being warm blooded and being cold blooded. Note: Scientists now prefer the new terms "endothermic" and "ecto-thermic," respectively. Birds and mammals are warm blooded and must maintain a constant body temperature. Yet, there are grada-tions of warm-bloodedness. Even within mammals, there is variation. Hibernating mammals, such as bears, can lower their body temper-ature during hibernation. We humans cannot do that. Being warm blooded has the advantage of immediate readiness to perform, but it comes with a high price tag. It is energy expensive to maintain a constant body temperature. We have to consume ten times the food of a cold-blooded animal in order to just stay alive. The danger of starvation is far greater for us.

Many dinosaurs were huge animals. Some were even the size of a Greyhound bus. When you are that big, it takes less heat genera-tion to maintain a constant body temperature than if you are small. You lose less heat to the environment too, because the skin area per unit weight decreases as you get bigger. So, big dinosaurs had the advantage of more constant body temperatures without the burden of a high metabolic rate to sustain it. One particular type of dinosaur was evolving into a truly warm-blooded animal. It was the dinosaur line that eventually became today's birds. Yes, not all dinosaurs went extinct because certain dinosaurs evolved into birds. Several species of feathered dinosaurs have been found in China. The original function of the feathers was for thermal insulation, not flight. Insulation makes sense only for animals that are attempting to preserve the heat their bodies have generated. The pterosaurs, those flying reptiles of the Mesozoic, were certainly warm-blooded animals too. They had a furry exterior in order to preserve body heat.

Speaking of birds, did you know that most, if not all, dinosaurs had gizzards. The gizzard works to digest food by grinding and grinding away. The dinosaurs, marine reptiles, and crocodiles of the Mesozoic used gizzards in the same manner as birds do today. However, you need to scale everything up. Those giant dinosaurs had giant gizzards to match their size. The gizzard stones were proportionally larger too. In fact, it is from these large stones (a.k.a. gastroliths), found within or near fossil dinosaur skeletons, that we know about their digestive methods.

Triassic, Jurassic, and Cretaceous

The Mesozoic era consisted of three periods: Triassic, Jurassic, and Cretaceous. The following table is a bird's-eye view of the Mesozoic era:

Table 3.3 Dinosaur Evolution in the Mesozoic Era

Time Span	Dinosaur Events	Plants Being Eaten
Late Triassic to Early Jurassic	Very slow return of life. Other reptilians dominate Dinosaurs first evolved	Mainly soft plants (Ferns, cycads, conifers and ginkgoes)
Mid and Late Jurassic	Radiation of dinosaur types Long necks, Brontosaurus, Stegosaurs	Tall plants with better defenses
Cretaceous	Radiation of dinosaur types Low feeders: beaked and armored dinosaurs, etc.	Low plants. Angiosperms evolve mid-Cretaceous

Triassic Period

The Triassic period began with a depopulated planet due to the great End-Permian extinction. The recovery of plants and animals progressed very slowly. The forty-million-year-long period was half over before the first dinosaur evolved and was nearly over when the first mammals evolved. Reptiles were evolving into various forms before the first dinosaurs appeared. The rhynchosaurs are one of them. These herbivorous reptiles resembled pigs in that their broad jaws and tusks served to uproot vegetation.

Globally, we see that the supercontinent of Pangaea was still intact. It first started to break up late in the Triassic with the Tethys Sea splitting it into northern and southern portions. Those new continents would be called Laurasia and Gondwana. The interior of Pangaea was an uncomfortably hot and dry place for life in the summers and quite cold in the winters, so life became concentrated on the periphery of the supercontinent, where climates were less extreme. The southern portion of Pangaea was over the South Pole and had frigid temperatures early on. This changed as warming increased during the Triassic. Great forests arose composed of cycads, conifers, and ginkgoes. These are all seed-bearing plants. Flowering plants did not evolve until the Cretaceous.

Dinosaurs first evolved in the latter part of the Triassic and radiated into the main types of dinosaurs we associate with the Jurassic period. Scientists sort them into two categories: ornithischia and saurischia. Ornithischia means "bird hip" dinosaurs and divides into subcategories, whereas saurischia means "lizard hip" and also divides into subcategories. The following table illustrates this. Images of these dinosaurs follow in the rest of the chapter

Table 3.4 Radiation of Dinosaur Types

Main division	Sub-division	Examples
Ornithischia	Thyreophora	Ankylosaurus, Stegosaurus
	Cerapods	Parasaurolophus, Triceratops
Saurischia	Sauropods	Brachiosaurus, Diplodocus
	Theropods	Allosaurus

Did the Triassic period end due to a meteorite impact? There are supporters of that hypothesis. Moreover, there actually was a large meteorite impact in the Triassic. It left a huge crater, sixty-two miles in diameter, in Quebec, Canada, and the meteorite itself may have been thirty-seven miles in diameter. The problem is that best estimates put it ten million years before the end of the Triassic period. Geological activity was intense toward the end of the Triassic. Rifting was well underway as Pangaea prepared to break up again.

Jurassic Period

The geological event of note during the Jurassic is the breaking up of the supercontinent Pangaea. The actual separation of new continents is preceded by the formation of rift valley and earthquake over millions of years. Clearly, it was the Jurassic Period that was the heyday of the dinosaurs. One reason is that competitors to dinosaur dominance were eliminated by the mass extinction at the end of the Triassic (i.e., the T/J extinction). Another reason was greener, lusher vegetation for them to eat. Dinosaurs essentially had the world to themselves. Hundreds of different species evolved as they found better and better ways of attaining food and avoiding predators. One line of defense against predators was better running speed and better sight and hearing. Another line of defense was armor plating, along with spikes and horns to discourage predators. Here are a few of the Jurassic Dinosaurs.

Figure 3.6 Some Dinosaurs of the Jurassic Period
Courtesy of Wikimedia Commons

Stegosaurus

The Stegosaurus dinosaurs are herbivores recognizable by their double row of bony plates. Those so-called armor plates may have not really been for defense. Instead, they may have been for regulating body temperature or for sexual attraction. Some species of stegosaurs had spikes protruding from the body. Those were probably for defense. Also note the spikes at the tip of the tail. Those were assuredly for defense.

Brachiosaurus

Sauropod dinosaurs such as brachiosaurus shown below were enormous compared to any land animal alive today. These herbivores ranged from 23 to 130 feet in length. Their massive size was itself a defense against predators. A blow from their tail or neck could send a smaller predator flying. Sauropod fossils have been found on almost every continent.

Allosaurus

Another suborder of dinosaur is the theropods, and Allosaurus is an example of one from the Jurassic. These predators were an average of twenty-eight feet in length, may have weighed a ton, and had a huge head with sharp teeth. Those small forelimbs are a curiosity to me. Did they have a useful purpose? Or were they vestigial?

Cretaceous Period

The breakup of Pangaea was well along in the Cretaceous with the continents of today emerging as independent landmasses. The oceans were rising, and the climate worldwide was warm and humid. More and more plants and animals of these new landmasses became different from each other.

Angiosperms Emerge

Animals are not the only life-form capable of defending themselves, plants are capable too. They can make their leaves toxic, grow defensive needles and thorns, or grow very tall. Dinosaurs had been leveling forests with their voracious appetites, and new kind of plant (i.e., angiosperms) appeared in the Cretaceous Period and exploded into new species. Why? Because these flowering plants had a big advantage over the slow-growing plants of the time. They were better at propagating their seeds, and they were more quickly taking root in bare areas. Today, we are surrounded by thousands of different kinds of flowering plants. It is satisfying to know that it was their superior adaptation to hungry dinosaurs that led to that species radiation.

Dinosaurs

The dinosaurs of the Cretaceous were copious and unique. For example, the horned dinosaurs made their debut. Triceratops is well known, but Centrosaurus, Styracosaurus, Pachyrhinosaurus, and Torosaurus are variations on the same theme. They all had huge bony head frills and horns. The Ankylosaurus were another herbivorous dinosaur of the Cretaceous identified by their triangular heads and armored/spiked bodies. They also could put up a fight with the hefty, bony club at the end of their tails. Ankylosaurus, Pinacosaurus, and Talarurus are examples. The duck-billed dinosaurs were still another herbivorous dinosaur and are called Hadrosaurids. They are typified by the flat, duckbill appearance of their snouts. Unlike the horned and armored dinosaurs, these dinosaurs were mainly bipeds. All these herbivores had good reason to be prepared for danger as the Cretaceous had its display of predatory dinosaurs as well. The Tyrannosaurs are discussed below.

Figure 3.11 Triceratops
Courtesy of Wikimedia Commons

Parasaurolophus

Notice how Parasaurolophus walks up on its toes? This dinosaur falls under the suborder called "ornithopods" for their birdlike feet. These dinosaurs ranged in size from six feet long to sixty-six feet long. They could move faster than previous dinosaurs and could digest a wider selection of vegetation. That was because they were the first dinosaur with true chewing teeth. The strange-looking head is unexplained. Perhaps, it helped amplify vocalizations or helped with thermoregulation or may have had to do with sex or species recognition.

Triceratops:

Triceratops means "three-horned face," and these herbivores first appeared only in the late Cretaceous, about sixty-eight million years ago. These dinosaurs reached lengths of thirty feet and weighed in at six to nine tons. Their skulls alone were over six feet long. These fellows kept an eye peeled for our next Cretaceous dinosaur, T-Rex.

Tyrannosaurus

Tyrannosaurus means "tyrant lizard," and the motion picture *Jurassic Park* made the terror of T-Rex real for millions of viewers. T-Rex was one of the largest of the predatory dinosaurs measuring forty feet in length and twelve feet tall at the hips. It was thought that T-Rex may have lost his crown when Giganotosaurus was discovered in 1995. However, it is more likely a tie. Another challenger to T-Rex is Anthrocanthrosaurus.

The Cenozoic Era

The Phanerozoic Eon does not end with the Mesozoic era; it also includes our era, the Cenozoic era. However, it is special for us because it is about us, the mammals. We pick up the thread in the next chapter and learn something about ancestors closer to us than dinosaurs.

Four

The Path to Our Primate Ancestors

The Chapter in a Nutshell

A global extinction occurred sixty-five million years ago when a six-mile-wide meteor struck the Earth on the Yucatan Peninsula. That extinction ended the Age of Dinosaurs (i.e., Mesozoic era) and brought in a new geological era (i.e., the Cenozoic) also known as the Age of Mammals. After the extinction, life returned in the form of new kinds of dominant animals, namely, mammals and birds. By ten million years after the impact, both mammals and birds had undergone considerable adaptive species radiation as they filled the empty ecological niches.

Mammals, being warm blooded, usually have fur coats to preserve their body heat. However, the name "mammal" comes from their method of feeding newborns. That is, providing milk via a mammary gland. The earliest mammals were called monotremes, and they laid eggs and suckled their young. Marsupials carried fetuses in a pouch, where they had attached themselves to a teat. Placentals carry the fetus to term in the uterus and feed them milk from a mammary gland after birth. Most of the mammals these days, including ourselves, are placentals.

Mammals can also be broken down into carnivores, cetaceans (i.e., whales), rodents, ungulates, and primates. One particular group of mammals, the primates, specialized in living in trees. Their adaptation included developing stereoscopic vision and grasping hands and feet. The first primates, the prosimians, persisted for millions of years but were replaced by Old-World Monkeys beginning thirty-three million years ago. Prosimians still exist on the island of Madagascar, off the east coast of Africa. In the Miocene epoch (23–5 mya), apes evolved from one group of monkey ancestors. Unlike monkeys, which have long tails and travel on top of branches, apes are tailless and swing beneath the branches.

Only a few ape species still exist in the world today, and they are endangered. In Asia, the agile gibbon lives with its mate in the tropical forest. Asia also has the orangutan. The male of this species can be much larger than the female. Orangutans usually live solitary lives and rarely descend from the trees. Africa has the more social gorillas and chimpanzees. Gorillas become too heavy for tree living when they reach adulthood. Groups of females and offspring are led by a dominant male known as a silverback. Chimps are smaller than gorillas and spend most of their time in the trees. The common chimp is male dominated, but the bonobos are female dominated and far less violent. Bonobos split off from the common chimp one million years ago.

Say Good-Bye to Dinosaurs, Marine Reptiles, and Pterosaurs

Creatures of the Cretaceous Period

Imagine taking a time machine journey back to sixty-five million years ago. On this trip, we will witness major planet-changing events, which will pave the way for our existence. Different creatures dominated the Earth back then. Large dinosaurs dominated the land, and the mammals cautiously stayed out of their way. Most of these mammals were squirrel sized and only became active at night when it was safer to scurry about. In the seas, there were no whales or dolphins yet. In their place, swam the sea monsters of the day, huge marine reptiles.

They were soon to perish in a catastrophic event, but perhaps, their days were numbered anyway. The fish of the seas had not stood still but were evolving into formidable opponents, at least with regards to eliminating the offspring of the marine reptiles. The lords of the air were flying reptiles called pterosaurs. Some of them reached the size of a small airplane. Yet as we look around this Cretaceous landscape, we can spot actual birds. Once a type of dinosaur that used feathers to stay warm, true birds evolved and put those feathers to a different use, namely flight.

Witnessing the Impact

We become aware of a new object in the sky. It is a very fast body on a path toward our planet. It is either a meteor or a comet, approximately ten miles across, becoming larger and larger in diameter as it nears the Earth. It has just slammed into the Yucatan Peninsula and instantly killed all living things for thousands of miles around it. It was a devastating event, but its destructive effect is just beginning. The shock wave generates enormous tsunamis, and their power changes the geological landscape and spreads death and destruction around the world. At the impact site itself, the impact's compression melts the rock to a red-hot mass, and the reaction hurls these molten, glowing bodies far into space. The heavier objects return to Earth soonest, whereas the smallest particles, such as dust, return much later or even go into orbit around the Earth. Forest fires are ignited by the red-hot rocks raining down over much of the Earth. And still, the destructiveness of the impact is not over. Even after the hellacious fires have burned their last, things do not return to normal. The Earth is cloaked in a persistent dust cloud that blocks the Sun. Plant life dies, and animals starve. By the time these devastating events were finally over, more than three quarters of all plant life and animal species on the face of the Earth had gone extinct.

Controversy and Resolution

How sure are we of this horrific scenario? The proof that an impact happened seems overwhelming. Not that there were no vocal opponents to the impact hypothesis. Many paleontologists believed that the dinosaurs had died out over many thousands of years and not in a catastrophic event. Many geologists intuitively favor explanations involving slow and gradual processes over explanations requiring sudden catastrophic events. Many opponents to the impact hypothesis felt that an adequate explanation for the extinction already existed. Extraordinarily heavy volcanic activity had already been underway in India for many millions of years during the time of the extinction. Enormous quantities of basalt flooded out in the Deccan Plateau of Western India. The lava beds that formed are called the Deccan Traps, and they are huge! They have a volume exceeding 240,000 cubic miles.

Many paleontologists had long accepted extreme volcanism as the cause of the extinction. Choking dust and lethal gases are by-products of these volcanoes. And so in the 1980s when the impact hypothesis was proposed, a great controversy developed in the scientific community, where one camp supported the volcanism hypothesis and the other camp favored the impact theory. Disputes like this are healthy for science because when reputations are on the line, involved scientists are working their hardest. The public follows the debates, and funding gets found to do the research needed.

The world had seen other global mass extinctions, so we can refer to this one by its name, "the K/T Extinction." The letter "K" signifies the last period of the Mesozoic, namely the Cretaceous. The letter "T" signifies the first period of the Cenozoic, namely the Tertiary Period. Now, let us present the arguments that persuaded most scientists to support the impact hypothesis for the K/T Extinction event: The most persuasive fact favoring the impact hypothesis is the presence of high levels of the rare element iridium in the stratum, where the extinction is observable, the K/T transition layer. Nobel Laureate Luis Alvarez, working with his son Walter, saw this iridium anomaly as evidence of an impact by an extraterrestrial body because it is the only explanation for such high levels of iridium. Moreover, additional testing of the K/T layer at one hundred places around the world produced the same iridium high readings and provided convincing evidence that the impact hypothesis was correct. In addition to iridium, high levels of charcoal were found at these sites. This evidence confirms the existence of worldwide fires ignited by red-hot rock fragments ejected from the impact site. "But where is this impact crater?" critics demanded. Luckily for the impact protagonists, it was eventually found. The earth is mainly covered by oceans and had the crater been in the deep-sea bottom; it might never have been found. Fortunately, it impacted on land. The crater is called the Chiculub Crater and is on the Yucatan Peninsula in Mexico. The asteroid or comet must have been at least six miles in diameter to generate the dust cloud containing the iridium, which was found at the K/T transition zone. The meteorite traveling toward us at forty thousand meters per hour would have left a crater 110 miles in diameter and twelve miles deep. It turns out that the size of the Chiculub Crater is a match for the meteorite, and the sixty-five-million-year-old age of its rocks is correct for the time of the K/T extinction event. Further evidence to support the Chiculub Crater as the impact site is the tidal wave debris and location and the size distribution of impact material relative to the crater. Shocked quartz and glass tektite beads at the site are further evidence of an impact. If you would like to dig deeper into this evidence, I suggest you read *Night Comes to the Cretaceous* by James Laurence Powell.

The Adaptive Radiation of the New Cenozoic Era

Until this world-changing event, our mammal ancestors lived in a world dominated by reptilian monsters. They included massive dinosaurs, soaring pterosaurs, and ferocious marine reptiles. These reptiles of land, air, and sea had long been kings that reigned over the other beasts, and now those kings were dead. After the K/T Extinction had run its course, the mammal's age-old predators were gone forever, and it was safe at last to come out into the daylight. And they did. Moreover, they found many empty ecological niches and began to fill them. And in the process of filling these empty niches, many new mammal species emerged. Over time, the process of natural selection will hone an animal or plant to optimally fill its new niche, thereby optimizing their chances for survival and proliferation. As this process occurs, the plants and animals gradually change into new distinct species. In the early epochs of the new Cenozoic era, many niches were being filled at once. When new species are forming at a rapid pace like this, scientists call this phenomenon "an adaptive radiation." Mammals were not the only ones undergoing adaptive radiation. Birds were undergoing an adaptive radiation of their own. Just think of the thousands of different species of birds in the world. We know now that birds actually evolved from feathered dinosaurs. So, in a sense, dinosaurs didn't totally go extinct. They still exist in the thousands of species of birds in the world.

The Cenozoic Era

Our human roots may have been begun with mammalian development in the Mesozoic era, but the story becomes a lot more interesting in the Cenozoic. Mammals, being warm blooded, fared better than the dinosaurs during the lethal conditions after the K/T impact. The planet was highly depopulated due to the extinction and abounded with empty ecological niches. The birds and mammals that survived the extinction would over generations begin to fill them. Table 4.1 provides a brief outline of how that progressed. The geological periods and epochs are indicated.

Table 4.1 Events of the Cenozoic Era

Period	Epoch	Dates	Features
Quaternary			
	Holocene	12000 ya - now	Ice age ends, Megafauna extinction, Modern man and the Agricultural revolution, Civilization and human dominance
	Pleistocene	2.5 mya - 12,000	Ice Ages, Megafauna evolve, The Homo lineage evolves,
Neogene			
	Pliocene	5 - 2.5 mya	A colder epoch than the previous Miocene. Grasslands and savanna increase and well as grazers. The Australopiths evolved
	Miocene	23 - 5 mya	Apes evolved and diversified, Grasslands spread at expense of forests
Paleogene			
	Oligocene	33 - 23 mya	Numerous kinds of mammals exist. As for primates, monkeys evolved,
	Eocene	55 - 33 mya	Whales evolving, as for primates, the prosimians proliferated.
	Paleocene	65 - 55 mya	Rain forests abound, Highest temperatures reached, radiation of mammals, primitive proto-primates

The Paleogene Period

The *Paleogene Period* was the warmest time in the Cenozoic era with a thermal maximum reached 55.8 million years ago. Climate change influences evolution, and the dominant factor in climate change was the breakup of the supercontinent Pangaea with newly formed continents adrift. India, for example, was on a collision course with the continent of Asia. Now, although the era began with high temperatures, humidity, and heavy forestation, a very-long-term cooling trend reversed that trend. The global cooling with its drier climates caused the breakup of the forests, the expansions of savannas, and the eventual dominance of grasslands.

New mammal species were developing in the Paleogene. Among them were dogs, cats, pigs, elephants, small horses, rhinos, and others. Birds were also diversifying at a similar fast pace, but we are more focused on the mammal's story.

The Paleogene Period is significant to our human story because the primates were first evolving. The Paleocene Period is divided into three epochs as shown in table 4.1, which are the Paleocene, the Eocene, and the Oligocene.

Land Mammals

Most of the mammals of the Paleogene period have gone extinct and would look strange to us if we saw one today. A reconstruction of the mammal "Carodnia vierai" is shown in the figure below. This mammal had evolved in the late Paleocene epoch and was the largest mammal of South America. It was about the size of a modern tapir. By the time of the Oligocene epoch, much larger mammals existed. There was a hornless rhino called indricotherium that roamed Eurasia. These long-legged, long-necked rhinos stood fifteen feet tall at the shoulder, were up to twenty-five feet long, and weighed fifteen to twenty tons.

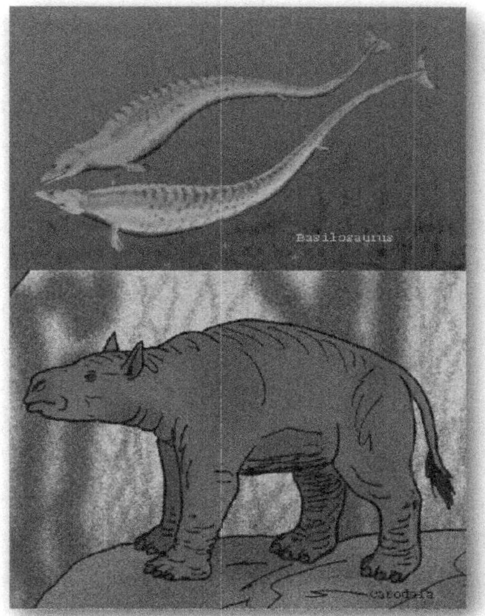

Figure 4.1 Some Creatures of the Early Cenozoic

Courtesy of Wikimedia Commons

Sea Mammals

There were many other quadrupedal mammals evolving to fill the empty niches left by the dinosaurs. Amazingly, a few of these quadrupeds actually gave up their land jobs and became sea creatures. Of course, that didn't happen overnight because evolution is a slow gradual process. One of them was Basilosaurus as shown in the figure 4.1. The name sounds like it describes a marine reptile because that is what they thought it was at first. Basilosaurus was actually a mammal of the Eocene epoch. Some of them measured eighty feet in length.

Flying Mammals

So, land mammals evolved into sea creatures in the Paleogene, did they also evolve into flying mammals? The answer is yes. Bats are 20 percent of all mammals and have radiated into over fourteen hundred living species. The oldest fossils of bats are dated between fifty and sixty million years ago. So, it appears they evolved quite early after the K/T extinction. There is still some controversy as to whether microbats evolved separately from megabats. Microbats are the smaller bats who eat insects, blood, fish, lizards, birds, and nectar. Megabats (a.k.a. flying foxes) are the large fruit-eating bats.

The oldest fossilized bats appear to lack echolocation traits, so flying or gliding using sight seems to have preceded flying by echolocation. This was learned from a very primitive fossilized bat called Onychonycteris. We should really consider bat evolution in the context of the concurrent bird adaptive radiation. Perhaps, bats could not compete with birds during the daylight hours and may even have had to hide away from predatory birds. However, with echolocation bats could fly during the darker hours, when most birds are grounded. One or two bats come out at dusk where I live, and their aerial dynamics are a thing to behold. They are catching flying insects at an amazing rate. They whiz around faster than the eye can follow. However, not all bats use echolocation, and the ones that do not are the fruit bats that live in the tropics. They are also the largest of the bats.

The Neogene Period

In the Neogene period, the continents were getting close to their current positions. India was now pushing so hard against Asia that the Himalaya Mountains began to rise, with a profound effect on global climate change. For example, monsoonal patterns in Asia affected glaciations in the northern hemisphere. The Neogene also saw the flourishing of Bovids (cattle, sheep, goats, antelope, and gazelles).

The Neogene Period is divided into two epochs as shown in table 4.1. They are the Miocene and the Pliocene. The Miocene epoch saw an increase in grasslands and other plants that we would recognize today. Increasing and diversifying grasslands led to a diversity of grazers. Examples are horses, rhinos, and hippos. Figure 4.3 is a reproduction of a mural in the Smithsonian Museum. It depicts the kinds of animals in North America during the Miocene. The Miocene epoch is important to human evolution because apes evolved from Old-World monkeys and diversified. However, the cooling trends with decreasing humidity caused the tropical forests to decline with drastic consequences for the apes. Many apes were forced to find a living in open woodlands and grasslands. Primitive bipedal apes, like the famous Lucy, evolved during this period.

Figure 4.2 Miocene Fauna of North America
Jay Matternes [Public domain], via Wikimedia Commons

The Pliocene epoch was a colder and more arid time than the preceding Miocene. Consequently, grasslands and savannahs spread at the expense of forests. Long-legged grazers evolve to eat the dominant vegetation. One geological change that influenced the climate was the linking up of North and South America. This event also had an effect on the plants and animals due to mixing with new competitors. It also affected marine species now isolated due to the new barrier. Meanwhile, India was colliding with the Asian continent causing the rising of the Himalaya Mountains. As the tropical forests declined, the tree-dwelling apes faced a problem of inadequate nourishment and the disappearance of safe terrain. Some apes evolved the bodies of bipedal walkers to better transverse the open woodlands in search of food. We call them the Australopiths, and they are the subject of the next chapter.

The Quaternary Period

The *Quaternary Period* saw the species radiation of bipedal apes in Africa. They had a brain size similar to that of a chimp. One of these species was ancestral to modern humans. The genus Homo is assigned to the fossil men who were taller, bigger brained, and thought to be ancestral to us. The Quaternary Period is divided into two epochs as shown in table 4.1. They are the Pleistocene and the Holocene.

The Pleistocene Epoch

The Pleistocene epoch was one of Ice Age animals because glaciers covered huge parts of the planet. Figure 4.3 shows an artist's rendering of some of them. We see a wooly rhinoceros, saber-toothed cats (i.e., Smilodon), wooly mammoths, and horses. This epoch began 1.8 million years ago, which was about the time that the first human is found in the fossil record. *Homo erectus* looked a lot more like us than any of the bipedal ancestors before it. The Pleistocene epoch ended 11,700 years ago, and by then our species, *Homo sapiens*, had occupied every continent on the planet except Antarctica.

Figure 4.3 Pleistocene Mammals
By Mauricio Antón [CC BY 2.5 (http://creativecommons.org/licenses/by/2.5)], via Wikimedia Commons.

The Holocene Epoch

The Holocene epoch is the time we live in now, and one of the things that make it significant is a reversal of the long cooling trend and a return to a warmer climate. The melting of the enormous glaciers caused the seas to rise by 115 feet, and continents relieved of their ice burden rose as much as 590 feet. Most of these geological events happened in the early Holocene. The Holocene also was when the megafauna such as mammoths, mastodons, wooly rhinoceros, and giant ground sloth went extinct. Scientists are divided on whether humans caused these extinctions or if other factors are responsible.

As major as these geological events appear, the most significant difference between the Holocene and all other previous periods is the transformational influence of one particular species, namely humans. Where herds of wild animals once roamed, cattle, sheep, and other domesticated animals have taken their place. Many species of animals have gone extinct from man's incursions, and many others are endangered. Forested have been leveled to make way for domesticated plants. Rivers have been dammed and rerouted. Roads, buildings, and parking lots displace natural habitat and alter the reflectivity of the planet's surface.

There are seven billion people on the planet now, and the population growth trend continues. Consider the issue of global warming. Will our addiction to fossil fuels convert our planet into another Venus with temperatures unsuitable for life? How about erosion of the ozone layer? This layer protects us from deadly radiation, and it is essential for life to exist. These problems and many others have life-or-death consequences, yet coordination and commitment to solve them is at times questionable.

The Emergence of Different Mammals

Mammals may well be the lead characters of the Cenozoic era. They certainly filled a diverse range of ecological niches, which were emptied by the K/T extinction. However, before we tell that story, we should first define what we mean by the term "mammal."

Definition of Mammals

The first thing to consider when defining a mammal is its warm-bloodedness. This trait is quite different compared to cold-blooded fish and reptiles; whose bodies change temperature to match the temperature of their surroundings. This reptilian system has a noticeable downside because their activity level is governed by their current body temperature. By contrast, mammal bodies are capable of maintaining a constant body temperature with the benefit of an optimal activity ability whenever it is needed. This ability may have lifesavings consequences if a predator suddenly appears, but it comes at a cost. The mammal is at much greater danger of starving than the cold-blooded reptile. Maintaining a constant body temperature means frequent eating to supply the necessary fuel to produce that body heat.

What else distinguishes mammals? Fur, of course! One effective way that mammals can retain that costly body heat is to insulate themselves with a coating of fur. Retain more heat, and you don't have to eat as often. Yes, it is true that not all mammals are fur covered. Some mammals, like the whales that live in the oceans, use an external layer of fat to insulate themselves. Yet whales originally evolved from land-dwelling, fur-covered ancestors. It turns out that blubber is a more effective insulation than fur if you live in cold water, and their bodies evolved to best match their new aquatic environment. Large animals like elephants are hairless too. Their great size preserves heat automatically. In fact, elephants need to worry more about overheating than being too cold. Let's not forget that their cousins, the mammoths and mastodons, were fur covered. They were large animals too, but they lived under very cold arctic conditions.

Some other things distinguish mammals, and the name gives one of them away. Mammals have mammary glands. They are able to suckle their young. Those baby mammals are warm blooded just like their parents and have to be fed regularly or they will quickly die. The mammary gland with its supply of nutritious milk solves the problem. Finally, there is one last trait that distinguishes a mammal, and that is diverse teeth. We mammals have a variety of different kinds of teeth in our mouths, which include incisors, canines, premolars, and molars.

Three Kinds of Mammals

Dinosaurs, like their reptilian ancestors, reproduced by laying eggs. In fact, fossil nests containing dinosaur eggs have been well documented. We mammals have a common ancestor with reptiles, so it isn't surprising that the earliest mammals also laid eggs. However, these monotremes did something new as well; they suckled their young. When did monotremes first appear? There is fossil evidence of them as early as the Triassic Period, which is the first period of the Age of Dinosaurs. Alas, the monotremes went the way of the Model T, and yet two species still exist in our time. They are the duckbilled platypus and the spiny anteater.

Natural selection favors the better model, and a better model than laying eggs is for the mother to keep the fetus with her physically. Marsupials do just that. Now when I think of marsupials, I immediately picture the opossum. They abound in the Los Angeles area and are often seen navigating the tops of our cinder block walls. Our dogs were playing with a very young opossum in the backyard one day. Its fur coat was coated with canine saliva, and it looked stunned. My wife stood over it and watched as it staggered a few feet, fell over sideways, and seemed to be dead. She had been played though, because after she went back in the house, it got up and ran to safety. What distinguishes a marsupial? It is their fetal-development system. A few weeks after conception, the tiny fetus crawls into its mother's pouch and attaches itself to a teat for the rest of its development. Australia, before man occupied it, had fauna, which was exclusively marsupial. It is still populated with marsupials, namely: koalas, Tasmanian devils, kangaroos, and wallabies.

Placentals are mammals that carry their fetus in the uterus to term. The fetus is nourished via the placenta. Most of the animals we know are placental mammals. Lions, tigers, elephants, cows, bison, monkeys, otters, seals, horses, pigs, goats, dogs, and cats are all placentals. Humans are placental animals too. Both marsupials and placentals evolved during the Mesozoic era, and both participated in the adaptive radiation of the Cenozoic era. Placentals seem to have won the contest for dominance. In the competitive struggle for survival, having the most and hardiest offspring is the key to winning.

The Road to Higher Intelligence

Mammals and birds were evolving during the Mesozoic but were restrained from radiating into many new species because the niches were already filled by reptiles, predominately dinosaurs. The K/T Extinction event changed the world markedly. Worldwide fires destroyed vegetation; massive dust clouds blocked the sun; and the cold, sunless period that followed killed plant life, which is dependent on photosynthesis. For those animals that survived the initial shock waves, it would take many weeks for the dust to settle and for the sunlight to break through again. Cold-blooded animals were especially vulnerable because their activity level is dependent on their body temperature, and that is determined by the temperature of their surroundings. The warm-blooded mammals and birds had the advantage of mobility and used it to find food and survive. Most mammals of the time had adapted to a nocturnal life in the Mesozoic, and that gave them a special advantage in the dark, cold world that resulted from the dust-shrouded cloud surrounding our planet.

Perhaps, the mass-extinction event only accelerated what was inevitable anyway; that warm-blooded animals would dominate over cold-blooded ones. The reason is that the warm-blooded animals are smarter, and higher intelligence wins out over brawn in the long run.

Brains work best when they operate continually at their ideal temperature, and that is why mammals were smarter than reptiles. Moreover, survival required that they be smarter. In fact, they were compelled by natural selection to be as smart as they could be. Remember that mammals need to eat regularly in order to maintain a constant body temperature, whereas reptiles need only one-tenth of that same food energy to survive. Mammals have to be continually opportunistic. One of the ways that they achieved higher intelligence is through the invention of the neocortex.

So, what is a neocortex? First, we can say what it is not. It is not a replacement for the reptilian brain, which controls basic functions like endocrine, immune, growth, and stress-response. The neocortex did not replace the reptilian brain in mammals; it added additional brain capacity to what existed. The neocortex does additional functions; it controls higher functions such as sensory perception, cognition, movement commands, spatial reasoning, and language. The neocortex is a part of the cerebral cortex. The neocortex is smooth in smaller mammals, but in the case of primates, it has grooves and ridges. The scientific name for the groove is the sulcus and for the ridge is the gyrus. Finally, the neocortex is composed of six thin layers of neural tissue.

Humans are primates as are the apes from whom we descended. The neocortex is especially vital to primates because they moved through a three-dimensional world of tree branches and empty space. We humans no longer live in trees nor swing through their branches, but that is the kind of brain we inherited. How have we adapted to the very different life of a human being? Our brains also have a property called neural plasticity, which is a fancy way of saying that our brains are flexible in being used for unforeseen functions. Generation after generation, our brains acquired new capabilities important to how we lived at the time. And so it is today that we can verbally communicate and develop languages. We can solve math and engineering problems. We can conceive of airplanes, spaceships, and other inventions before they exist.

Primates

Primates are a type of mammal, whose bodies and minds are highly adapted to a life in the trees. For example, primates can escape their predators by moving faster and more securely through the trees than can their predators. Primates, unlike ground dwellers, move through a three-dimensional world including long leaps from branch to branch. Consequently, their eyes are not on the sides of their heads like most mammals but have moved to the front of their faces so that they can look straight ahead with both eyes and better assess distances. We call this feature stereoscopic vision, and we humans can thank our ape ancestors for having it ourselves. We use it too. It is a great advantage when driving a car and even more important for flying an airplane.

Primate Evolution

After the K/T extinction sixty-five million years ago, life was slow to rebound at first. Yet within ten million years, birds and mammals filled the empty ecological niches with new species at an ever-increasing rate. During the first epoch of the Cenozoic (i.e., the Paleocene), primitive prosimians came into existence, but it was during the next epoch (i.e., the Eocene) that prosimians having modern-day characteristics evolved. Not only did the eyes move forward to provide stereoscopic vision, but the foramen magnum (i.e., where the spine attaches to the skull) moved to raise the head to a forward-looking position. Brains became larger. There was an adaptive radiation of prosimians in the Eocene with numerous new species. However, their populations plummeted once monkeys evolved and outcompeted them.

Aside from the physical differences between primates and other mammals is the matter of intelligence. Primates generally have bigger brains per unit body weight than other mammals. Moreover, these primate brains have more cortex volume and a greater density of neurons. The result is that primates can think deeper and faster than other mammals. These abilities also have importance in social interactions, and primates do have complex social interactions as researchers like Jane Goodall have reported.

Prosimians

The earliest primates were squirrel-like but over time evolved stereoscopic vision and other primate abilities. The first true primates are called prosimians. The heyday for prosimians ended when the monkeys evolved millions of years later. You can still see prosimians in the wild if you visit the island of Madagascar off the east coast of Africa. Or you can see them in the zoo. They are the lemurs, lorises, and tarsiers. They have foxlike faces with large eyes, forward-looking heads, and grasping hands. In the Oligocene (33–23 mya), monkeys evolved and had an adaptive radiation of their own. As monkey populations increased, prosimians declined. Prosimians survive in Madagascar today because monkeys never got to this island.

Figure 4.4 Lemurs

Courtesy of Wikipedia Commons by Adrian Pingstone (Own work) (Public domain), via Wikimedia Commons.

Ape Evolution

DNA evidence indicates that apes and Old-World monkeys had a common ancestry twenty-five to thirty million years ago, but until recently we didn't have fossil evidence to verify that early date. Now, new specimens of fragments of ape jaws and teeth found in Tanzania have improved the picture taking ape ancestry back to twenty-five million years ago. We used to say that apes first evolved in the Miocene, but now it looks as though they first evolved in the Oligocene epoch. The Miocene was the high point of ape existence though. The climate was ideal for their tropical forest existence, and they occupied the most territory that they ever would during this epoch.

DNA evidence allows us to track the divergence of ape species over time. The gibbons and orangutans of Asia diverged from the African apes about twelve to eighteen million years ago. The African apes include the gorillas and chimpanzees. By the way, please don't use the terms "apes" and "monkeys" interchangeably. Apes differ from monkeys by being smarter, tailless, and by traveling through a forest by swinging under the branches of trees (i.e., brachiation). Monkeys have tails and only run atop the branches. Apes can recognize themselves in a mirror whereas monkeys think they see another monkey in the mirror. Now, wouldn't you rather claim that more perceptive ape as your relative over the unperceptive monkey? Moreover, humans don't have tails and have a similar free rotation in the shoulders, which allows us to swing under branches too.

Asian Apes

Gibbons

I love to watch gibbons perform because they are excellent acrobats. That might be true because gibbons are the most primitive of the living apes. They are actually closer to monkeys in many ways. Consequently, gibbons are grouped separately from the great apes. They do not make nests like other apes, and the males and females are similar in size. Gibbons are found in the tropical rain forests of Bangladesh, India, China, and Indonesia, where they form male-female pairs, which often last a lifetime.

Orangutans

Orangutans are bigger than gibbons and are distinguished from the other apes by their auburn hair. They are found only in the rain forests of Borneo and Sumatra, and not surprisingly, the two species of orangutan are called the Bornean orangutan and the Sumatran orangutan. Orangutans are most arboreal of the great apes, and that is probably because it is safer for them high in the trees. To venture lower may make them a meal for the hungry tigers and crocodiles at ground level.

One curious thing about orangutans is that the males come in two different sizes. The larger version is much bigger than the female, whereas the smaller version is only a little bigger. Since a typical female orangutan is three feet nine inches tall and weighs 82 lb., that larger male dwarfs her with his four-foot six-inch height and 165 lb. weight. Females only permit the large-version male to mate with them. However, the smaller male does not simply give up. He has been known to rape a female if no one is around to stop him.

African Apes

Gorillas

Gorillas are large ground-dwelling apes and mainly eat plants. There are two species of gorilla, the Eastern gorilla and the Western gorilla. Gorilla troops are ruled by one powerful male, and that mature male maintains a harem of his exclusive females and their offspring. Any other mature male that could be competition is driven out of the troop. Males weigh between three hundred and four hundred pounds in the wild and up to six hundred pounds in captivity. The males are much larger than the females. Moreover, the males use that great size to defend their troops against carnivores, humans, or other male gorillas. Despite their fierce appearance, gorillas are basically gentle animals.

Figure 4.5 Gorillas
By David Stanley from Nanaimo, Canada (Gorillas) [CC BY 2.0 (http://creativecommons.org/licenses/by/2.0)], via Wikimedia Commons.

Chimpanzees

Chimpanzees are closely related to gorillas yet have a totally different lifestyle and society. Chimps live in large social groups of both males and females. Usually the group is led by a dominant male, who maintains order during disputes. However, unlike the male gorilla, this alpha male chimp lives with his male rivals on a daily basis. He has to use political skill as well as physical prowess to stay dominant. Female chimps live a political life as well. Their relative position in the group may be inherited. The ranking position of a parent can determine the status of the child. This social complexity indicates the intelligence of chimps.

Physically, a chimp is adapted for movement under tree branches. Its arms are longer than its legs. In fact, a chimp's stretched-out-arms span is 50 percent greater than that of a human's span. Chimps have been seen fabricating and using simple tools. They use rocks to break open nuts and use sticks to fish for termites. Although they are unable to speak, chimps have been taught to communicate in sign language by humans. However, they seem to be limited to two-word sentences.

Unlike the gorillas with their harems ruled by a mature male, chimps live in social groups of both males and females, and many males have sex with a female in heat. Consequently, male chimps have much larger testicles than do male gorillas. Instead of the males battling to pass their genes on to the next generation, their sperms do the battling. Hence, the larger testicles to produce more sperms. Despite their social tendencies and friendly appearance, chimps can be violently aggressive and territorial. They are much stronger than us too. They can shimmy up a forty-foot tree in seconds.

Bonobos

Figure 4.6 A Bonobo
By Fanny Schertzer (Own work) [GFDL (http://www.gnu.org/copyleft/fdl.html) or CC BY 3.0 (http://creativecommons.org/licenses/by/3.0)], via Wikimedia Commons.

The Congo River poses an impassable barrier between the common chimp and their sister species, the bonobo. The bonobos live south of the river, and the common chimp lives north of the river. At one time, they were of a common species. However, now after a million years of being separated, they have very different cultures. Females have a higher social status than is observed with chimps. The bonobos don't fight as much as chimps. Instead, they resolve most social problems with sexual favors, not violence. It is a free-love society, where sex is part of daily life. Moreover, it is not restricted to male-female sex. The females often have sexual relations with other females by rubbing their genitalia together. So, why are they so much mellower than chimps? It may be the easier life. It is easier to find food south of the river that may account for the milder dispositions of the bonobos.

To the untrained eye, they look like the same animal. However, the bonobos are smaller than the chimps. Whereas male chimps average 3.9 feet in height, Bonobos are slightly shorter. They are also thinner than chimps and have longer limbs. Both chimps and bonobos are knuckle-walkers. That is, they walk on all fours by clenching their fists and support themselves on their knuckles. Both can also walk upright albeit unnaturally. They might walk upright while carrying something with their hands. Bonobos are seen walking upright more often than chimps.

The Australopiths

Scientists tell us that humans are descended from African apes, which lived five to seven million years ago. Many people greet that news with skepticism because humans are not only physically different from apes but able to speak, write, read, and invent things. In the next chapter, we will see that climatic change made life very hard for the tree-dwelling apes. Some of them traversed the open spaces in the diminishing forests in search of sustenance and became bipedal walkers. We call these bipedal apes the Australopithecines or simply the Australopiths. Their story is next.

Five

The Australopiths

The Chapter in a Nutshell

How did we humans come into existence? Genetic evidence points to chimpanzees as our closest living relative, although we are related to the gorilla as well. In fact, it took some ingenious scientific effort to establish that the chimp was closer to us than the gorilla. That same DNA evidence estimates that we humans and chimps were together as one species five to seven million years ago. Our ancestors split off from them due to a prolonged global cooling trend and the need to survive in a changing environment, where ground travel was necessary to find new food sources. Fossil evidence shows that there were apelike creatures walking on two legs nearly five million years ago. Actually, we have found several species of them in Africa. They are similar enough to be called by the same genus, namely *Australopithecus*. We therefore title the chapter "The Australopiths." The most famous of the Australopiths is Lucy. This 3.4-million-year-old nearly complete skeleton was a significant find. The good fortune continued with the find of several other partial skeletons, collectively called the First Family. Lucy's species has been dubbed *Australopithecus afarensis* to honor the Afar region of Ethiopia where she was found. This chapter also looks at *A. africanus, A. anamensis, A. garhi*, and the newly found

A. sediba. We also mention 4.4-million-year-old *Ardipithecus ramidus*, which is not strictly an Australopith but may have been an ancestor to them. The Australopiths developed the human ability to walk on two legs but retained their small apelike brains. One of the Australopith species evolved into the genus "Homo" and eventually us. It may be one of the ones discussed, or perhaps it has yet to be discovered.

Background

Genetic comparisons between African apes and modern man indicate that we had a common ancestor some five to seven million years ago. Where does that estimate come from? It comes from genetic evidence, which is to say that scientists have compared the DNA of apes and humans. They examine the DNA sequences for mutational markers. Those markers, which are shared by both species, occurred during the time we humans were together with chimps in a common species. However, those markers that are different in today's humans and chimps obviously occurred after the two species split apart. The markers are dated by a process employing computers, and the time of species split can be estimated.

These theoretical projections are impressive, but what does the fossil evidence say? Actually, we do have some fossil evidence to consider. The fossil record does show that an upright walking apelike creature was evolving between five and two million years ago. Paleoanthropologists call them Australopithecines, or Australopiths for short. We humans are bipedal walkers too, so the Australopiths seem to represent an intermediary stage in the evolutionary transition from ape to man. Examination of these ancient fossil bones reveals a pattern. Essentially, the Australopiths were undergoing extensive skeletal changes in order to become better and better upright, bipedal walkers. Their tree-based chimp-like cousins stayed quadrupedal and evolved to become knuckle-walkers when they moved on the ground.

This chapter is devoted to better understanding the Australopiths. We will be looking at the evidence and interpretations of the experts who specialize in paleoanthropology. Occasionally, they have been very fortunate in discovering nearly complete skeletons and finding more than one individual, from which to define a new species. The more evidence that I see, the more confidence I have in the extrapolations made about these creatures. By the way, we need a word other than "creatures" to apply to these bipedal walkers, and the word "Hominin" is the traditional one. The word "Hominid" is also seen in the literature and used by some authors. Hominin infers the bones belonged to an individual on the path to becoming human, whereas Hominid could include an ape not on that path. Hominin is a subset of hominid, so if one is not sure what the bones represent, hominid is a safer choice.

Global temperatures were at an all-time high some fifty-five million years ago and continued falling until twenty thousand years ago. The climate of the Miocene was highly favorable to ape populations and their ideal territory (i.e., extensive tropical forests). However, the temperatures continued falling during the Pliocene, and as more and more moisture became locked up in arctic ice, humidity declined as well as temperature. The tropical forests with their abundance of fruits and nuts shrank in size as grasslands increased. A cooling planet was converting extensive tropical forest, which were ideal for tree-dwelling apes, into open woodlands, savannahs, and even deserts. Numerous apes faced starvation if they could not find food somewhere. Moreover, the problem continually worsened generation after generation. Upright-walking apes (i.e., the Australopiths) evolved during the Pliocene, and we know this from their fossil record. The Australopiths left evidence of their existence in the limestone caves of South Africa and the eroding rocks of Northeast Africa. They may have been in other parts of Africa too, but conditions for fossil preservation were not as good in those places.

The Species Genetically Closest to Us

Man and Apes Are Genetically Related

Skeptics may say that apes exist, humans exist, and Australopiths may have existed, but that doesn't mean any of those three creatures are related. Why should I believe that they are? Scientists answer that evolution is what ties them together. In fact, when Charles Darwin wrote *The Origin of the Species by Means of Natural Selection* in 1859, scientists of the day had already begun to speculate about human evolution. At the time, Darwin discussed the evolution of plants and animals in his book but avoided applying his evolutionary theory to humans. He felt that he had already rocked the boat enough by declaring that species do change to a society that had long believed the opposite. So, he left the subject of human evolution for a later day. Remember that the church wielded tremendous power and influence over what was deemed acceptable thinking back in Darwin's day, and they had already determined that the species had been created in a perfect form and never changed thereafter. In other words, species were immutable.

Darwin finally did confront the human evolution issue, and in 1871, when he wrote a book titled *The Descent of Man and Selection in Relation to Sex*" where he took on this controversial subject. He expressed the view that man most likely evolved from the great apes in Africa. This was a brilliant insight it turns out, but it was also an unpopular idea at the time. Today's scientists have investigated the man/ape genetic relationship question by comparing the DNA of humans with the DNA of all the existing apes. The gibbons and orangutans of Asia were ranked most distantly related, whereas the gorillas and chimpanzees of Africa were most closely related to man than any other animal.

The Hominoid Trichotomy

Until recently, genetic scientists were unable to resolve the question of which of the living apes was closest to us humans. Was it the chimp or the gorilla? It boils down to a comparison of their related genes. Now, it turns out that mammals have essentially the same genes for the same purposes. These similar genes are called "homologous genes." The homologous genes of apes and man are particularly similar to each other. In order to compare two species, their gene sequences are scanned for mutations, which have left their mark over the generations, Mutations common to both species, probably occurred before the species had split apart, whereas any mutations, unique to only one of the species, probably occurred after the two species split apart. Thus, it can be estimated when the species split apart.

You are probably familiar with family trees, where one's ancestors are graphically depicted to show the lines of descent. A species tree is a similar depiction, except the connected boxes contain species, not individuals. Using the species-comparison information just mentioned, a species tree can be constructed. In the case of the three species (humans, chimps, and gorillas), all three have descended from a common ape ancestor. The species tree should show us which of the three species split off first. The remaining two species are the ones most closely related.

Now, numerous studies were done by comparing particular homologous genes of man, chimp, and gorilla in an effort to construct a species tree and resolve the question of genetic closeness of these three species. Alas, the results were inconclusive! Some gene studies showed chimp and man closest, some showed gorilla and man closest, and still others showed chimp and gorilla closest. The dilemma became known as the "Hominoid Trichotomy." In time it was realized that these conflicting results reflect reality. We are descended from both chimp and gorilla ancestors, so it is natural to have gene trees indicating that. In order to determine which particular one is genetically closest to us, we have to examine all the genes. The conflict was finally resolved by sequencing the entire genomes of these three species and applying statistical analysis to the results. We humans are most closely related to the chimpanzee, but there are also genetic contributions from gorillas and even more-distant apes.

The Missing Link

Now back to Darwin's day, scientists began to speculate about the description of a hypothetical missing link, a creature that was halfway intermediate between the tree-dwelling ape ancestors and modern man. It wasn't as though they had no data with which to begin. After all, Neanderthal fossils had already been found in Europe. The problem was that, the scientific community had yet to acknowledge that Neanderthals were a separate species. So, the missing link's hypothetical description was arrived at by logical deduction. They reasoned that it had to have a brain volume about midway between ape (400 cc) and man (1350 cc), and it should use its hands to make tools. They further reasoned that an upright posture and walking on two legs came later. The conclusion was that the bigger brain had to develop first, so that there would be a reason for the missing link to assume an upright posture and mode of locomotion. Sometimes, a preconceived expectation can prevent you from recognizing the truth when it confronts you. We will discuss this next.

Bipedal Apes

Raymond Dart and the Taung Child

In 1924 Raymond Dart discovered fossil evidence of a hominin in South Africa, but it certainly did not fit the expected description of the missing link and because of that mismatch, his discovery was dismissed for decades. Raymond Dart was an accomplished man. He was born and educated in Australia. He studied geology and zoology and later medicine and surgery. He served in the Australian army as a captain and a medic during World War I. We find him in Johannesburg, South Africa, in 1922, where he was a professor of anatomy at the University of Witwatersrand. Two years later, his life changed when he was presented with two crates of fossils recovered from a limestone quarry. Imagine his reaction when he recognized one of the fossils as being a skull of a hominin child. Among anthropologists today, this landmark skull is known as the "Taung Child." Dart named the species, *Australopithecus africanus*. This was the first evidence that Charles Darwin had been correct fifty-three years previous in his prediction that man evolved in Africa. Unfortunately, Dart's discovery and interpretation were not taken seriously by the scientific establishment of the time. The cranial capacity of the postulated missing link was supposed to be about 850–900 cc, but the value for these hominins was half of that. It is easy for critics to attack when a species is represented by a single skull. The fact that it was a child's skull weakens the case further.

It was not until years later that hundreds of additional fossils of *Australopithecus africanus* were discovered at places like Sterkfontein, Kromdraai, and Swartkrans in South Africa. This overwhelming evidence erased former doubts, and the species was confirmed to be hominin. These fossils have been dated at around three million years old. One of the key discoverers was Robert Broom, who was a Scottish doctor working in South Africa. In addition to being a doctor, he also happened to be a paleontologist, so his verification of Dart's discoveries carried substantial weight. One of the hominin individuals was a nearly complete skeleton. Examination of these bones showed that that these hominins walked upright on two legs. Like modern apes, they had short legs and long arms, which suggests that they spend time in the trees too. Half a century later, a different species of Australopith was discovered in Ethiopia. That story is next.

Donald Johanson and Lucy

Donald Johanson's name is recognizable to anyone who is remotely interested in paleoanthropology. Unlike many of the other great hominin discoverers, he is an American, born in Chicago, Illinois, to Swedish immigrant parents, although he later moved to Hartford, Connecticut. He began college with a chemistry degree in mind but later switched to anthropology. He focused on chimpanzee dentition as his doctoral thesis. This specialization served him well! In fact, it took him to Europe to examine the excellent collection of chimp teeth there and to South Africa to examine their Australopithecine teeth.

In 1973 he headed an American team in a joint effort with the French to hunt for fossils in Ethiopia. His counterpart on the French team was Maurice Taieb, a geologist and paleoanthropologist. Johanson worried that he would come up empty, but he actually found the femur and tibia of an ancient hominin. Here are the circumstances leading to the find: The team was working near Hadar, a village on the southern edge of the Afar Triangle. Three different geological plates meet in this area. This part of Africa has been a geologically active area for millions of years. The Great Rift Valley is a result of that seismic activity. The site of the hominin leg bones had been dated by Taieb and turned out to be about three million years old. That made Johanson's find the first of a kind. Examination of the fossils showed that the owner of these ancient legs was capable of walking upright proficiently. Experts deduced that fact from the angle of the knee joint.

The following year, Johanson's luck got even better. He discovered Lucy, perhaps the most famous ancient hominin of all time. Her name came from a Beatle's song, which was popular in the day, "Lucy in the Sky with Diamonds." This small, female bipedal ape had a great deal of her skeleton intact. She had lived long enough for her wisdom teeth to come in. She was small brained, although her skull was not complete enough to get a measurement of her cranial capacity. During their third field trip to Hadar, Johanson and team hit the jackpot of fossil finds. They found the so-called First Family. It was all filmed by a National Geographic's photographer, David Bell, who happened to be visiting. He actually filmed one of the most remarkable hominin fossil finds ever. In all, thirteen individuals were found, and that included men, women, and children.

In 2006 a different team of anthropologists found more fossils of Lucy's kind at Dikka, which is across the river from Hadar. Zeresenay Alemseged of the Max Planck Institute and his team found a very well-preserved skeleton of a girl about three-year-old. The skeletal find consists of a whole skull, the entire torso, parts of the upper and lower limbs. This child answers an important question, "What did Lucy's face look like?"

Johanson knew that he had a very important discovery handed to him, and he felt that it was very important to do a careful, complete, and professional job of interpreting the fossil evidence, which had been collected. So, he asked the American paleoanthropologist, Tim White, to assist in the analysis and interpretation. After exhaustive analysis, they decided that they had discovered a new species of hominin that lived between three and four million years ago, walked bipedally with feet much like ours, probably spent some of their time in trees, had an apelike face, and was small brained (380–450 cc). Their brains were only slightly bigger than that of modern chimps, and there was no evidence that these hominins made stone tools. Johanson and White named the new species *Australopithecus afarensis*, after the Afar region in which it was discovered. They further proposed that Lucy and her kin were probably ancestral to the Homo lineage.

Lucy and her kin must have got around because now we are in Tanzania. While Johanson was busy in Ethiopia, Mary Leakey was busy here, and she found something really unique. She and her team found humanlike footprints, which have been dated at 3.6 million years old. Mary is no longer with us, but besides being the wife of and fellow fossil hunter of Louis Leakey, she is famous in her own right. Their numerous and important contributions have earned them the title of "the first family of paleontology." These prehistoric footprints were preserved in volcanic ash and indicate that three individuals were walking together. This famous Laetoli site contains the first direct evidence that hominins walked bipedally. Fossils of *Australopithecus afarensis* have been found nearby the site, and the footprints match the contours of this hominin's feet. So, it is reasonable to conclude that Lucy's kin left those footprints. The site also contains thousands of footprints of animals of that ancient time; they include elephants, giraffes, rhinos, and several extinct animals.

Important hominin finds continued to be found in Northeast Africa. There were so many that it seemed foolish to waste one's time searching for hominin fossils anywhere else. However, Lee Berger disagreed with that wisdom and changed paleoanthropology as a result of his convictions. His story is next.

Lee Berger and the Malapa Fossils

Professor Lee Berger is a paleoanthropologist with the University of the Witwatersrand, in Johannesburg, South Africa. Even though so many important finds have been in northeast Africa, he has argued that South Africa is where the significant human evolution took place. We already discussed the fact that South Africa was where Raymond Dart discovered the hominin species, *Australopithecus africanus*. Well in the year 2008, South Africa got the spotlight focused back on it again. For it was on August 15 of that year, that Lee Berger's nine-year-old son, Matthew, found the fossilized clavicle of an extinct hominin embedded in a rock. That discovery would lead to a new species of hominin in South Africa.

Berger had been exploring limestone caves in an area called the Cradle of Humankind World Heritage Site for years, but on this occasion, he brought along his son and their dog. He told young Matthew to go find a fossil, and to his surprise, Matthew quickly found a rock with a hominin clavicle sticking out of it and presented it to his dad. Berger turned the rock over and felt a sudden wave of excitement. There sticking out of the rock was a hominin mandible, still containing a canine tooth in it. When Berger returned again to the site with fellow scientists, they found nothing further for quite a while. Then Berger spotted hominin bones on another rock within the pit. When he lifted the rock, two hominin teeth fell into his hand. Soon, other fossils were found. Some fossils were hominins, and some fossils were animals, such as saber-tooth cats, mongooses, and antelopes. This pit, known as Malapa Cave, was much deeper when these victims were alive two million years previous. We know that age to be accurate because the site was dated using radioisotopes (i.e., Uranium-Lead). We also know how these individuals died. They were the unfortunate victims of fatal falls. This is not speculation; it has been proven to be a fact. Analysis of the stress lines in the bones was consistent with a hard fall.

Berger has declared that these hominins comprise a new species, which he calls *Australopithecus sediba*. The word "sediba" means natural spring in the Sotho language. The species is based upon the six individuals discovered at the Malapa Fossil Site. They include a juvenile male, an adult female, an adult male, and three infants. The juvenile male, designated MH1, was four feet two inches tall and had a brain capacity of 420 cc. His mandible and teeth were humanlike, but his brain size was too small to include him in the Homo lineage. Note: The genus *Homo*, which is different from *Australopithecus*, is the subject of the next chapter. The hominin fossils were in excellent condition, and many skeletal bones were found. For example, the finger bones were in all found and could be assembled to complete the whole hands of the individuals.

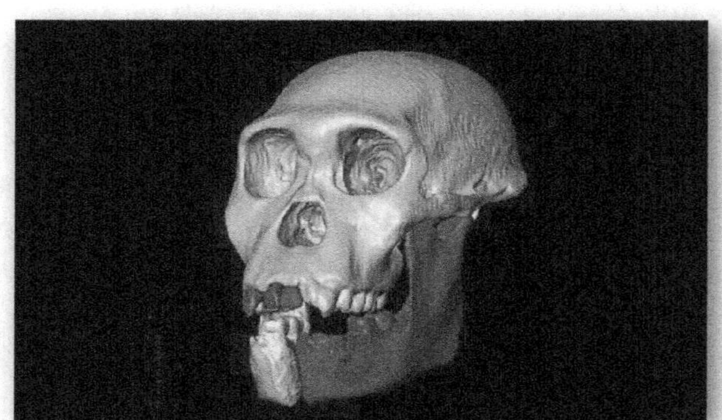

Figure 5.1 Skull of *Australopithecus sediba*
By Profberger (Own work) [CC BY-SA 3.0 (http://creativecommons.org/licenses/by-sa/3.0) or GFDL (http://www.gnu.org/copyleft/fdl.html)], via Wikimedia Commons.

The *Australopithecus sediba* fossils had a cranial capacity range of 420–450 cc, which is apelike, yet these individuals had long legs, which is a human characteristic. In fact, the characteristics of this species seem to be a mosaic of apelike and humanlike traits. The hand, for example, is surprisingly humanlike and one with a precision grip, which might have been used for toolmaking. Additional individuals have been found at the site, which will reinforce and add to the description of this new hominid species.

Meave Leakey and Allia Bay Fossils

Meave Leakey is one of the famous Leakey family, which has given so much to the study of early man. The tradition began with Louis B. Leakey (August 7, 1903–October 1, 1972). Louis was born in Kenya, the son of British missionaries. He had seen stone tools of ancient man during his youthful adventures and became determined to prove to a skeptical world that Africa was the cradle of humanity. With his dedicated wife, Mary (February 6, 1913–December 9, 1996), and small three sons (Jonathon, Richard, and Philip) helping him, Louis succeeded in that goal. One of the sons, Richard, was to match his father in great accomplishments. Richard, like his father, married a woman who became a skilled paleoanthropologist herself. Richard married Meave, and it is Meave who is credited with the discovery of four-million-year-old *Australopithecus anamensis*.

Meave Leakey obtained her PhD in zoology in 2004 and has an honorary D.Sc. in paleontology. Archaeologist Alan Walker is also credited with the discovery of *A. anamensis*. Walker, a professor of biology and biologic anthropology, has also worked on important hominin finds with Maeve's famous husband, Richard Leakey. Together, Meave Leakey and Alan Walker excavated a site near Allia Bay in Kenya. Which Allan Morton had worked thirty years previously. Morton had found fragments of a specimen protruding from the hillside but came to no firm conclusions. Leakey and Walker found additional fragments at the site, including a complete lower jawbone. Although the shape of the jawbone was apelike in appearance, the teeth were intermediate between ape and man. Examination of the postcranial bones caused them to conclude that this hominin walked bipedally.

Australopithecus anamensis, as the new species was called, is believed to be ancestral to Lucy's clan (i.e., *Australopithecus afarensis*).

So far, twenty individuals of this species have been identified from nearly one hundred specimens. Bones include upper and lower jaws, cranial pieces, parts of a tibia, and a humerus. The body mass of these Australopiths is estimated to range between 66 and 112 lb. Isotope dating of volcanic ash sampled from above and below the specimens, bracket the age of these fossils to between 4.1 and 4.2 million years. The species may have existed from 4.2 million to 3.9 million years ago. In 2006 a new *Australopithecus anamensis* find was made in Ethiopia, extending their range beyond Kenya.

Tim White and Ardi

Tim White, professor at the University of California at Berkeley, is another American paleoanthropologist. He has worked with many of the prominent paleoanthropologists but has made his own important discoveries as well. While working on his PhD, he did fieldwork with Richard Leakey in Koobi Fora, Kenya. He also worked with Mary Leakey at Laetoli in Tanzania. And as you might recall, he helped Don Johanson analyze the First Family fossils. As for his own enterprises, White discovered an extinct subspecies of *Homo sapiens* in Ethiopia in 1997. The fossils of three individuals were dated at 154,000 to 160,000 years old.

The discovery of interest to us in this chapter is of a fossil hominid much older. I am referring to his discovery of a 4.4-million-year old hominid near the Awash River in Ethiopia in 1994. The nearly complete female skeleton was deemed to be a new species, namely, *Ardipithecus ramidus*. White felt that the species was too primitive to be included in the genus *Australopithecus* and assigned it a new genus, namely, *Ardipithecus*. However, I am including this species in the chapter on Australopiths because it seems to belong here. After all, Ardi is only a few hundred thousand years older than *A. anamensis,* and they lived in the same geographical area. If more fossils from this time and region are found, we may gain a better picture of the relationship between these bipedal ape species.

Ardi was a three-foot eleven-inch tall 110 lb. female, when she was alive some 4.4 million years ago. Tim White's team found her fossils in the Afar Depression of Ethiopia in 1992. Her bones were only forty-six miles distant from those of Lucy. We can say a lot about Ardi because much of her fossilized skeleton was preserved through this incredible span of time. She had a small brain (300–350 cc) similar to that of a bonobo. Her body was adapted for tree climbing as indicated by her long arms and curved fingers. However, she was also adapted to bipedal walking. She had evolved a foot shaped more like a human than a chimp, except that the big toe (i.e., hallux) was opposed to the other four toes. Moreover, her pelvis was more like a human pelvis than an ape pelvis. She had thin enamel on her teeth indicating a soft food diet (e.g., fruit). Her canines were not as long as ape canines. Some of these observations about Ardi and her kind came from examination of the fossils of nine individuals discovered in 1999 by a different team led by Sileshi Semaw.

Tim White and *Australopithecus garhi*

In 1996 Tim White teamed with Ethiopian paleontologist Berhane Asfaw. As we time travel back to that year, we find the two of them excavating in the Middle Awash of Ethiopia's Afar Depression. They have just discovered a new species in deposits 2.5 million years old, which they name *Australopithecus garhi*. The word "garhi" means "surprise" in Afar language. The sparsity of fossils prevents us from saying very much about this new species. Yet there is one significant thing that makes them special. These are the only Australopith species with evidence that they made and used stone tools. Moreover, these tools were used to butcher animals as attested to by tool marks on the bones. This fact causes some anthropologists to believe that *Australopithecus garhi* was ancestral to the *Homo* lineage.

Another significant fact gained from the fossils is that *A. garhi* had longer legs than *A. afarensis* and that means longer strides and more efficient walking. One of the fossil bones was a femur, and it was measurably longer than that of *A. afarensis*. We also know that *A. garhi* had large back teeth and a brain capacity of 450 cc.

An Overview of the Australopiths

What do we see when we look at the Australopiths collectively? Table 5.1 summarizes some of the features of these various ape-men. The locations, where these Australopith fossils have been found, seem to either to be Kenya/Ethiopia or South Africa. This is not to say that these ape-men only lived there. The Kenya/Ethiopia region is favorable to finding fossils due to its eroding landscape, whereas South Africa fossils are usually found in limestone caves. As for the time span for the Australopiths, it seems to begin a little more than four million years ago and seems to end around two million years ago. A span of over two million years would seem long enough to see some brain-size expansion, yet these bipedal apes have brain sizes similar to the chimpanzees of today. *A. garhi* is of special interest to us. Although there is a limited amount of fossil material, the long femur and animal bones plus stone tools found with the fossils suggest that *A. garhi* might have been a precursor to man.

Table 5.1 Comparison of the Man-Apes of Africa

Name	**Location**	**Age,** million years	**Brain size,** **cc**	**Comments**
Ardi*	Ethiopia	4.4	300-350	Transitional biped
anamensis	Kenya/Ethiopia	4.2 to 3.9		
afarensis	Kenya/Ethiopia	3.6 to 3.0	380-430	Lucy and others
africanus	South Africa	3.3 to 2.1	300-500	
garhi	Ethiopia	2.5	450	Tool-maker
sediba	South Africa	2.0	420	A recent find

*Ardi (*Ardipithecus ramidus*) is not classified as an Australopith but seems to be close to being one.

Physical Adaptations in the Australopiths

Walking, skipping, and running are as natural to us as breathing. We take these abilities for granted. In fact, we are not only excellent walkers, but we can run fast and for long periods. Considering that we evolved from tree-dwelling ape ancestors, a great deal of bodily change had to occur for us to be able to function so well. Chimps and gorillas rarely walk upright because the position is uncomfortable for them. Instead, they use a quadrupedal stance to move around on the ground. What they do is called knuckle walking. It works well for them and gives them the stability of a four-legged animal. Primarily, ape bodies are adapted for movement through the three-dimensional world of tree branches. Not only can they grab onto branches with their long arms and curved fingers, but their feet have grasping ability too. The big toe (i.e., hallux) is opposed to the other four toes. We, on the other hand, have five inline toes. We have feet adapted for walking and gave up the grasping foot in exchange.

Ardipithecus ramidus, who lived over four million years ago, was a remarkable fossil find because a nearly complete skeleton was recovered. From that skeleton, we can deduct that this hominid was capable of bipedal walking, but it had not developed feet with five inline toes. Its big toe was still opposed to the other four toes. *Australopithecus anamensis* followed Ardi in time and had a more advanced foot with all toes in line. *Australopithecus afarensis* (Lucy's species) lived a little closer in time to us. It lived in the same part of Africa as *anamensis* and *ramidus* but 3.4 million years ago. Lucy's kind also had five inline toes.

The Australopiths are sometimes referred to as bipedal apes. They are still regarded as apes because even though they mastered walking, their skulls were still apelike and because they still depended on trees for protection from predators. Their long arms and curved fingers were retained in order to quickly take refuge in the trees. Where big cats and hyenas hunt at night, it is very dangerous to sleep on the ground. The Australopiths probably continued in the ape tradition of making nests in the trees at night. Where their lifestyle differed from those tree-dwelling apes was during the daylight activities. They had to spend more time on the ground searching for the traditional foods, which were becoming increasingly hard to find due to climate change.

An ape's pelvic bones are not ideally shaped for walking and an upright posture, where it is necessary to support the weight of intestines. A more bowl-shaped pelvis evolved in the bipedal hominins. Apes do not need to support their organs with their pelvis and consequently have broader hips. Figure 5.2 shows a comparison of modern ape and human skeletons. Notice the difference in the pelvises. The difference in arm lengths is impossible to miss.

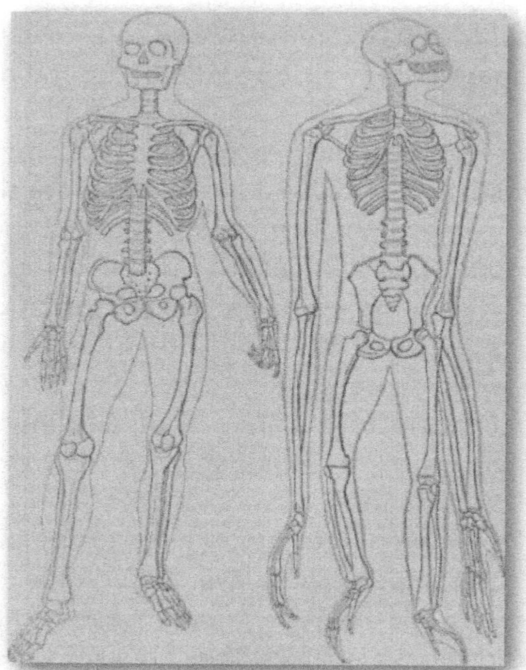

Figure 5.2 Comparison of Human and Ape Skeletons
Courtesy of Wikimedia Commons, by Internet Archive Book Images—https//www.flickr.com.

The Australopiths were in a state of evolutionary change, which optimized their ability to get around well using bipedal walking. Their ever-changing environment rewarded such adaptations. The favorite ape foods of fruits and nuts were getting increasingly hard to find. So, those individuals, who were able to get around and find it, passed their genes forward and those who failed to do this, did not. Consequently, the Australopiths evolved bodies suited to the changing environment. That list of traits included inline toes, bowl-shaped pelvis, and a skull/spine orientation optimal for an upright posture. Now, some fossil species of hominins are only represented by skulls or skull fragments, and yet it is still possible to tell whether they stood upright or not. There is a hole in the bottom of the skull, where the spine enters, and it is called the "foramen magnum." It is positioned too far back in apes to best balance the weight of the skull if they were to assume an erect stance. The foramen magnum is more centrally located in bipedal walkers. So it is clear that extensive changes have occurred in our transition from tree-adapted apes to bipedal apes and to *Homo erectus* and finally to modern humans.

Intelligence of the Australopiths

On one hand, the Australopiths show no appreciable increase in brain size and from that fact, some might conclude that they were no smarter than apes are today. However, recent evidence suggests that brain size is not nearly as important as brain content. Most of that evidence comes from fossil hominins of the Homo lineage, so that discussion will appear in the following chapter. However, there is also evidence from the brain of *Australopithecus africanus*, which is applicable to this chapter. The arguments are beautifully presented in Dean Falk's book *The Fossil Chronicles*. I will only briefly touch on them here. Dr. Dean Falk is an American anthropologist, who has devoted her long career to understanding the role of the brain in human evolution. Over one-third of her book reexamines the efforts of Professor Dart and the Taung child endocast. An endocast occurs naturally when the brain cavity of the skull fills up with sediments and become fossilized. It turns out that Dart had valuable information in a scientific paper about the brain of the Taung child, but it never surfaced. The paper was not accepted for publication in a scientific journal due to the politics of the time (i.e., 1920s). Moreover, the information stayed obscure until Dean Falk carefully studied those writings, compared Dart's conclusions with her own observations of the brain endocast, and realized the great significance of Dart's paper. Falk's book reveals these important discoveries at long last. Endocasts capture the shape of the brain's surface, which contains valuable information about the brain's evolution and the functions of that brain. After extensive analysis, Dart had concluded that *Australopithecus africanus* had superior mental abilities compared to living apes. He wrote that "their eyes saw, their ears heard and their hands handled objects with greater meaning and to fuller purpose than the corresponding organs in recent apes." Dean Falk reached similar conclusions from her examination of the endocast. It would appear that the evolutionary mental progress of Australopiths is going to be a topic of interest in the years ahead.

Let's do a thought experiment together and imagine ourselves back in the days of the Australopiths. Natural selection was in overdrive during these times, and using one's brains had to be an important part of their adaptation process. Unlike the tree-dwelling apes from which they descended, the Australopiths no longer had an environment in which they were superior to their enemies. Instead, they were forced to spend a greater percentage of their waking hours on the ground searching for nourishment as their traditional foods diminished or disappeared altogether. Whenever they were on the ground, they were at greater risk of predator attacks. It seems reasonable to assume that enhanced alertness, better cooperation in spotting danger, and signaling that danger to their companions were traits selected for. Their brain evolution would favor all these traits. Dart reached similar conclusions about evolutionary changes in the brain from his examination of the Taung child's endocast.

The Australopiths also had to plan ahead and have an escape plan in mind at all times. Getting up a tree as fast as possible was one of those lifesaving options. Of course, all prey animals must have an escape strategy, and it usually has become instinctive for them. However, the Australopiths were at a disadvantage compared with the typical grazing animals in that bipedal walking was new to them. They simply could not outrun a big cat like most grazers can. Nor could they fight with the predator as some big animals can. This is not saying they weren't tough. I'd wager that the average Australopith could have easily killed a modern human. They were, as modern chimps are, many times stronger than we are. They could also open their mouths much wider and bite down harder. We have lost that ability. Now, if a number of hominins could coordinate their defense and maintain the discipline to not break and run, they could bluff the big cats. I have seen incredible videos of hunters with bow and arrows taking pieces of a lion kill away from the cats as they were eating it. That takes more courage than I have. The ongoing battle between lions and hyenas seems to be determined by relative numbers. Hyenas kill lions if they outnumber them enough. These animals make determinations all the time on whether a battle is worth engaging in or not. So, it is conceivable that Australopiths learned to weigh their chances in a similar way.

It seems that they did find a way to successfully adapt to the more dangerous life at ground level. After all, they survived for well over two million years. And we might just leave the topic there if it weren't for the discovery of *Australopithecus garhi*. This species is clearly associated with stone toolmaking and using those tools for scavenging meat. This reflects an intelligence not thought to be yet developed in Australopiths. It has long been assumed that hunting originated with the Homo lineage, but recent finds push the first hunting evidence back to over three million years ago. At a site in Dikka, Ethiopia, stone tool marks on bones for flesh and marrow removal date back to 3.4 million years. It would appear that Australopiths were hunters or scavengers as well as human ancestors. If that is true, we are going to have to reexamine how we rate their intelligence level.

Six

The Homo Lineage and the Dawn of Intelligent Life

The Chapter in a Nutshell

Around 2.5–3.0 million years ago, climatic changes forced the Australopiths to disappear and new types of hominins to replace them. Fossils from this period indicate that hominins employed two quite different strategies to survive in the changing climate. The first approach led to the genus *Paranthropus,* and their survival strategy was to eat whatever edible plants they could find. That meant roots, grasses, and more fibrous vegetation and heavy chewing in order to break down these fibrous plants. The result was fossils of these hominins, recognizable by their large teeth, large jaws, and bony structures to support their massive chewing muscles. Their strategy worked for about one million years, but they disappeared from the fossil record after that.

The second group are called the Homo genus, and their survival strategy was to add meat to their diet whenever possible. They probably scavenged dead animals at first, but they eventually evolved into hunters in their own right. They became increasing adept at knapping, a technique of making stone tools and weapons. Nature had not provided them with claws or fangs, so ingenuity was their way to become top predator. Meat is a highly nutritious food, so it is part of the reason that the Homo lineage evolved into a very different kind of hominin.

One change was their phenomenal increase in brain size. Today, our brain-to-body-size ratio exceeds anything seen in the animal kingdom. Other changes include radical physiological changes to more efficiently cool the body. A profusion of sweat glands developed in the skin, and the bodily fur was shed to make sweating more efficient. Venous changes to enhance brain cooling also occurred. Changes to improve running ability include long legs, shorter arms, barrel chest, and the cooling adaptations previously mentioned.

The chapter discusses the different species in the Homo lineage, including *Homo habilis, Homo erectus, Homo heidelbergensis, Homo neanderthalensis,* and others. *Homo erectus* is a species that first migrated from Africa to Europe and Asia. They were able to tame and use fire and eventually used it for cooking. *Homo florensiensis,* a.k.a. the Hobbit, is an unusual species, which is discussed at length. Its small brain and stature were not expected in a toolmaker and hunter. Also, *Homo naledi* is an exciting new find full of mystery.

The Human-Origin Story

Where did our species, *Homo sapiens,* come from? That depends on how far back in time that you wish to go. The answer is that we descended from African apes if you want to go back five to seven million years. Although accurate, that answer is unsettling because we are very different from apes in both appearance and abilities. Apes have fur, long arms, short legs, and small brains, and they live in trees. We, on the other hand, have big brains, long legs, short arms, no fur, and have little interest in climbing a tree. If we go back about two to three million years, then the answer is that *Homo sapiens* ultimately descended from one of the Australopith species. That answer is a little more believable. After all, the Australopiths did walk upright and like us, managed to survive in a variety of climates. Yet we can point out differences too. They had much smaller brains, longer arms, and shorter legs than we do. It is highly doubtful that they could carry on a conversation, and they probably fled up a tree at the slightest alarm. All right then, let's go back in time again, but this time only 1.5 million or so years. This time, startling changes are taking place in our ancestor's bones and their behavior. The hominin fossils appearing in this time frame are much more like us. These fossils have skull with bigger brain cavities, and skeletons with longer legs, shorter arms, and vertical rib cages. They are so obviously different from their bipedal ape ancestors, that we can no longer call them by the genus *Australopithecus*; we need a new genus name. That name is *Homo*, and we modern humans, who are part of that heritage, are called *Homo sapiens*.

The Dawn of the Homo Lineage

It seems highly likely that our genus Homo evolved from one of the Australopith species, and paleoanthropologists fervently debate the question of which one it is. Homo lineage species start appearing about 2.5 million years ago, but they still have primitive characteristics, and the fossil evidence is sparse for this period. So, why are we so eager to assign a new genus to them? The answer is that paleoanthropologists suddenly have more evidence to consider than just fossil bones. They have stone tools, which have been fashioned for cutting. Certain kinds of stones, like flint and obsidian, break when properly struck to produce a sharp cutting edge. The process of this kind of toolmaking is called knapping. A crude type of stone tool (i.e., Oldowan culture) appears 2.5–3 million years ago, but it is followed by advanced stone-tool cultures through human evolutionary time. The so-called Stone Age is an immense span of time in human history. Even when the Copper Age, Bronze Age, and Iron Age appeared in more recent times, much of the world still remained dependent on stone tools.

What Gave Rise to the Homo Lineage?

The Homo lineage represents a sequence of species, which were undergoing radical evolutionary change. So, there must have been a powerful stressor acting on these hominins to affect such radical change. What was it? We know that the Australopiths evolved from tree-dwelling apes into efficient bipedal walkers out of necessity. The planet was cooling, and the fruit-laden tropical forests, so ideal for apes, were breaking up into open woodlands and grasslands. Thus, an upright, bipedal method of movement on the ground was an adaptive solution to the need to find nourishment by searching the open woodlands.

During the time that the Homo lineage emerged, that cooling trend had made survival even more difficult. Savannahs and deserts began replacing many of the open woodlands. The fruits and nuts, which once were abundant, were becoming increasingly difficult to find. The evolving Homo lineage found a fortuitous solution to the diminishing food supply; they supplemented their diet with meat. This solution was fortuitous because meat is rich in fat, protein, and other nutrients that the body needs. However, acquiring meat required a new set of mental skills and physical traits.

Stone toolmaking was an essential skill needed in the acquisition of meat. Evidence abounds, which shows that the Homo lineage evolved into proficient hunters. However, it is highly probable that they began their meat eating as scavengers of fresh kills. They may have watched from safety as big cats brought down a prey animal, ate their fill, and secured the carcass for a later meal. In a team effort, they posted lookouts as others approached the secreted carcass. Here we see that those stone-cutting tools may have spelled life or death for these early scavengers. With the cutting implements, they could cut off large chunks of the carcasses and transport them to safety before angry cats discovered their loss. There were probably numerous times when they barely made it to safety. As for their ability to run, that was also a vital evolutionary development. I will say more on that later.

Paranthropus, a Parallel Evolutionary Branch

Homo was not the only new genus to emerge from the changing environment 2.5 million years ago. The traditional foods of tropical forests had either disappeared entirely or were rarely seen. Some Australopiths responded to the problem by evolving into a new genus, which ate whatever they could find even if it was very hard to chew, for example, fibrous vegetation, roots, or grasses. Natural selection rewarded those individuals, who could process such difficult food. Remember that their tree-dwelling ape ancestors had adapted to soft fruits, nuts, insects, and other easily processed foods. So, it was a major adaptation for their bipedal descendants to process coarser foods. It meant developing larger teeth, thicker enamel, stronger chewing muscles, and larger stronger jaws and skulls. And those are the traits that distinguish this new genus, called *Paranthropus*. Not all anthropologists use that term; some retain the *Australopithecus* genus name and distinguish them as "robust Australopiths." Not only are these robust hominins distinguished by large teeth, jaws, and chewing muscles, but they also may have a sagittal crest. In other words, they had a bony ridge on the top of their skulls, where their strong chewing muscles attached. Several authors tell us that Paranthropus reminds them in many ways as gorillas. Perhaps, the association comes from the fact that both survive on low-nutrition foods, which require spending most of the day chewing.

Table 6.1 Description of Paranthropus Species

Species Name	When Alive, mya	Brain capacity And Body size	Comments
Paranthropus boisei	2.1 to 1.1	500 to 550 cc, 4 ft. 6 in tall (male) 4 ft. 1 in (female)	More massive face and molars than *P. robustus*. Referred to as "Nutcracker Man". Found in Northeast Africa.
Paranthropus robustus	2.3 to 1.5	410 to 530 cc, 4 ft. tall (male) 3 ft. 2 in (female)	A body like A. africanus, but massive face and molars, found solely in South Africa. 130 individuals found in the cave at Swartkrans.
Paranthropus aethiopicus	2.7 to 2.5	410 cc, Height unknown	NE. Africa, Black skull specimen, massive muggy face, Little fossil evidence for this species.

Table 6.1 will give you an idea of the different species assigned to this genus. A point of interest is that *Paranthropus* lived at the same

time as Homo. Moreover, these *Paranthropus* hominins inhabited many of the same areas at the same time as the *Homo* hominins. It is highly likely that they interacted with each other. Another point of interest is the short life-span of *P. robustus*. Out of 130 individuals found in the Swartkrans cave, very few exceeded seventeen years of age. This certainly indicates that they lived under challenging circumstances. A third point of interest is that *P. bosei*, with the most radical chewing adaptation, ate a different kind of food than *P. robustus*, which consisted of grass like plants. Perhaps, this is due to them living in different regions of Africa. *P. bosei* occupied northeast Africa, whereas *P. robustus* occupied southern Africa.

Paranthropus evolved from the Australopiths, which preceded them. They survived by physically adapting to chew and digest the plant life available in the altered landscape. However, after about one million years ago, they no longer appear in the fossil record. They lived concurrently with species from the *Homo* lineage for about one million years before vanishing, so that indicates they were successful for quite some time. Perhaps, there is something about the relationship between *Homo* and *Paranthropus* that led to their eventual demise. We examine the *Homo* genus next.

Homo habilis and Other Early Homo Hominins

Louis Leakey (August 7, 1903–October 1, 1972) was born in Kenya to British missionaries. He became a world-famous anthropologist and succeeded in his life's goal of proving that man first evolved in Africa. Although man's origin in Africa is accepted as fact today, it was a ridiculed concept back in Leakey's day. It was especially unpopular in Europe to consider an African origin to humanity, and yet, Leakey was not basing his claim on guesswork. He had personally seen ancient stone tools during his African adventures, and he was convinced that human ancestors had made them. Together with Mary Nichols, who later became his wife, he toiled in Olduvai

Gorge, Tanzania, for years to find proof of that African ancestor. Sometime between 1960 and 1963, Louis and Mary discovered a new hominin species, which Louis believed made the numerous ancient stone tools found nearby. He named the species *Homo habilis*, which means handy man.

Some anthropologists differ with Louis and would assign the fossils to the genus *Australopithecus* instead. However, this ambiguity shows that evolutionary transitions occur only gradually. There is never a sudden change from one definite form to another definite form. Evolution is a gradual incremental process that happens over numerous generations. For those favoring the *Homo* genus assignment to *habilis*, the following arguments apply: The brain size of *Homo habilis* ranges from 550 to 687 cc. An average value of 640 cc is reported. This is 50 percent larger than the typical Australopiths brain. It was the larger brain and the stone tools, which gave Leakey the confidence to declare this species to be of the *Homo* lineage.

Another contender for being the first of the *Homo* species is *Homo rudolfensis*. Again, there is controversy as to whether to classify this species as *Homo* or *Australopithecus*. Louis died the year that Skull 1470 was found, but it was his son Richard who headed the team that found it. The skull was discovered at Koobi Fora, Kenya, in 1972. Skull 1470 has been dated at 1.9 million years old, and it has a brain capacity of seven hundred cubic centimeters. Richard's ability to fossil hunt was unfortunately ended in 1993 due to a crash of his personal airplane and the need to amputate his crushed legs. However, the Leakey dynasty didn't end there because Richard's wife, Meave, continued the family tradition of hominin-fossil discovery. In 2012 Meave's team added to the *Homo rudolfensis* fossil collection by finding a face and two jawbones also in Kenya. The 2012 fossils were dated at about two million years old. The differences between the skulls of *H. rudolfensis* and *H. habilis* are that the skull of *H. rudolfensis* is rounder and is less muggy (i.e., prognathous) and has smaller teeth than *H. Habilis*. Those traits are more like those of modern humans. It would be nice to have more fossils show up from this important time when the evolutionary path to modern humans truly began.

Homo erectus

While there is debate on whether some of the early *Homo* species should be instead designated as *Australopithecus,* there is no debate about classifying the more recent *Homo erectus* fossils in the genus *Homo*. That is because they don't look all that different from us. In fact, the main differences between *Homo erectus* and modern man are that *Homo erectus* had smaller brains, lacked a jutting chin, had prominent bony eyebrow ridges, and a few other characteristics. Otherwise, they are a lot like us with long legs, hairless bodies, and rounded skulls. Moreover, they look that way over a million and a half year ago!

Kinds of *Homo erectus*

The species designation, *Homo erectus,* is often used as a catchall for several different species of the same general description. *Homo erectus* is the name of a chrono species, that is, a species that is changing with time. It is also changing with location for fossils of *Homo erectus* have been found on three continents. Let us examine them, continent by continent.

Africa

The African *Homo erectus* is also called *Homo ergaster.* A mandible in South Africa was found in 1949 and another in Kenya in 1975, but surpassing both of these finds is a nearly complete skeleton of *Homo ergaster* found at Lake Turkana, Kenya, in 1984. The skeleton was found by Kamoya Kimeu and Alan Walker. Both of these fossil hunters have worked closely with the Leakeys over the years. In fact, Richard was actively involved in excavating the site, while Meave painstakingly assembled the skull fragments, which had been scattered by the expanding roots of a tree. This famous skeleton is known as KNM-WT 15000 or more commonly, Turkana Boy. It is dated at 1.6 million years old. Although he was between eight and twelve years old, according to latest estimates, he was tall for his age. He would have likely grown to five feet three inches in height had he lived to adulthood. He had a big brain (880 cc) for a hominin living back then. These larger brains infer something about infant care for *Homo ergaster*: Based upon the maximum size head that a can be delivered at birth, we must conclude that most of their brain growth must have happened after birth as is the case with humans today. And like today's humans, their infants had to be cared for many months as that postnatal brain growth occurred.

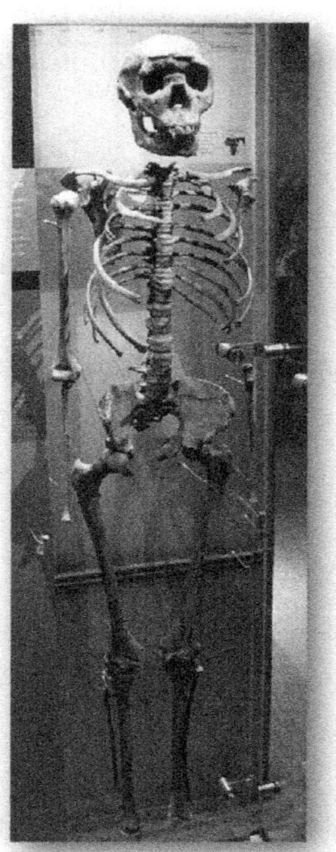

Figure 6.1 Turkana Boy or KNM-WT 15000
Courtesy of Claire Houck from New York City, USA (Turkana Boy) [CC BY-SA 2.0 (http://creativecommons.org/licenses/by-sa/2.0)], via Wikimedia Commons.

ASIA

The Asian *Homo erectus* is just called *Homo erectus*. It is generally believed that *Homo erectus* evolved in Africa and then migrated to Eurasia. There is fossil evidence of *Homo erectus* in Georgia, India, Sri Lanka, China, and Indonesia. The Dutch surgeon Eugene Dubois (January 28, 1858–December 16, 1940) discovered a *Homo erectus* fossil skullcap and femur in Indonesia long before anyone in the Leakey family even thought of fossil hunting. In 1891 Dubois discovered the fossil known as Trinil 2 and named the species *Pithecanthropus erectus*. That name was later changed to *Homo erectus,* which means "upright man." More *Homo erectus* fossils were found in Zhoukoudian, China. In all, two hundred fossils representing more than forty individuals have been found. These fossils have been dated at between 500,000 and 300,000 years old. In 2009 more fossils were found dated at roughly 750,000 years old.

EUROPE

The European *Homo erectus is called Homo erectus georgicus,* and these fossils are far more primitive than their African or Asian counterparts. The discovery site of these 1.8-million-year-old fossils was in Dmanisi, Georgia, and the discoverer was Professor David Lordkipanidze, a Georgian anthropologist. Four skulls were excavated in 1991, and a very complete skull was found in 2005. The brain size of these skulls measured about six hundred cubic centimeters. Postcranial bones from four individuals were also found. The individuals were about four feet tall. In addition, seventy-three stone tools were found useful for cutting and chopping, and twenty-four animal-bone fragments were also found. The discoveries in paleoanthropology often have unexpected results, as does this one. With a brain capacity of only six hundred cubic centimeters and an age of 1.8 million years, this species is more like some Australopiths than like typical Homo lineage species. If we didn't have the proof in front of our eyes, we wouldn't have expected such early hominins to be able to make the migration from Africa to Europe, or perhaps even make stone tools and butcher animals with them. Obviously, this discovery has had a huge impact on our thinking regarding Homo erectus. Perhaps, *Homo habilis, Homo rudolfensis*, and others should be classified as *Homo erectus* in light of this discovery.

The Hobbits

It has been proposed that *Homo florensiensis* is a descendant of *Homo erectus* based upon its anatomical similarities. For example, both species are chinless and have brow ridges and receding foreheads. If this new species is a direct descendant of *Homo erectus*, then a dwarf version of *Homo erectus* has existed until very recent times. If you are not familiar with *Homo florensiensis*, the fossil bones and tools of these so-called Hobbits were found in Liang Bua limestone cave on Flores Island in 2003. This 220-mile-long island is one of the thousands of Indonesian Islands. If you look at a map, Flores is just a short way down the chain of islands from Java, where *Homo erectus* fossils were discovered by Dubois in 1891. It is possible that they had made this journey to Flores Island long ago.

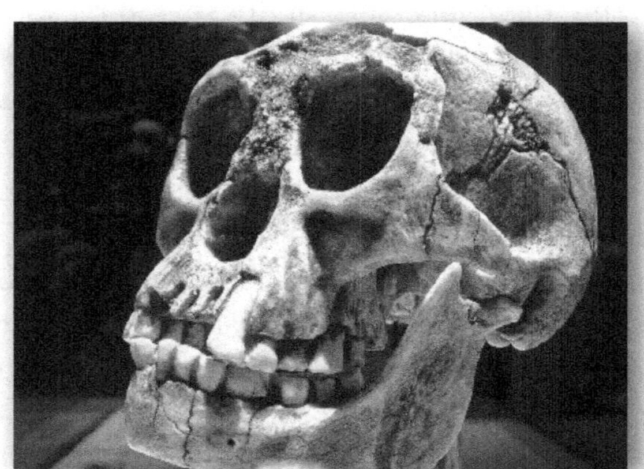

Figure 6.2 Skull of the Hobbit

By Gerbil (Own work) [CC BY-SA 3.0 (http://creativecommons.org/licenses/by-sa/3.0) or GFDL (http://www.gnu.org/copyleft/fdl.html)], via Wikimedia Commons.

Partial skeletons of nine individuals were found deep below the surface of Liang Bua cave, but only one skull was found. It was the skull from an adult female, and it was quite small. These three-foot six-inch tall hominins used stone weapons to hunt dwarf elephants (i.e., stegodonts) and giant rats on the island. The stegodonts bones add up to forty-seven individuals, but the cautious Hobbits selected infants and newborn to kill and not the more dangerous adults. They also occasionally killed the Komodo dragons, which in turn hunted them.

Now, it is unexpected that *Homo florensiensis*, with such a small brain, could fashion stone tools and use them to hunt. There is even evidence that Hobbits used fire. These abilities were thought to originate with *Homo erectus* and his or her enlarged brain. Granted that the hobbit had a small body to go with the small brain, but at a brain capacity of four hundred cubic centimeters, this species seems below the usual threshold for being considered a member of the Homo genus. How old were these little people when alive? Most of the reports, which I have seen, proclaim that the Hobbits were alive around 18,000 years ago. However, other sources date the hobbit fossils in a range from 100,000 to 60,000 years ago and date the stone tools found with the fossils in Liang Bua cave to a range between 190,000 and 50,000 years ago. Their occupation of the island may predate those dates because at Wolo Sege, similar stone tools were found that date to over one million years old.

There was a lot of skepticism when these findings were announced, and this species has only been found on this one island. And yet, the original interpretations have survived. That is, they are a legitimate new species of hominin. Moreover, a total of nine fossilized individuals were found, which rules out the likelihood of a freak occurrence. The lesson learned from these findings is that intelligence and skills like making stone tools are not as related to brain size as we had come to think. Perhaps it is the composition of the brain cells and genes, and not the number of them that is most important.

Regarding Hobbit's miniature size, there may have been an evolutionary reason for their size reduction. Islands are known for their tendency to accelerate evolutionary change. The stegodonts on Flores Island were only four feet tall, perhaps half the size of the animal normally. If being of smaller size has survival and reproduction advantage to a species, then it will get smaller and smaller over many generations until it has maximized its adaptation to its environment. We have no comparative data for hominins on islands, but it may be that the island effect is why Hobbits are so small. Another possibility is that they were that small when they first occupied the island.

The Hobbit story became a lot more interesting thanks to Dean Falk's recent book *The Fossil Chronicles*. Professor Falk, who is an expert in human-brain evolution, was called in to resolve several disputes concerning the Hobbit. Several scientists doubted that Hobbit was actually a new hominin species and that such a small-brained hominin could fashion stone tools, control fire, and hunt animals. Some of them believed that Flores Man was not a new species, but a modern human with a microcephalic brain. Professor Falk compared the Hobbit skull CT scan to CT scans of numerous other hominin skulls and several microcephalic skulls. Photographs of the resulting virtual images are presented in her book. She concluded that Hobbit was normal and not microcephalic. Moreover, Hobbits did not have the thin bones and very low intelligence exhibited by microcephalics.

She also did her own assessment of Hobbit's origins and concluded that although having some erectus-like features, Hobbit had many unique and more primitive features and could not have been descended from *Homo erectus*. Hobbit's massive feet are one such unique feature. Hobbit's short legs and long arms are unlike *Homo erectus*. Its brain size relative to body size was more like an Australopith than like a human ancestor. An apelike wrist is another reason that Falk thought Hobbit resembled an Australopith in many ways. On the other hand,

Hobbit had humanlike features too like wide temporal lobes, convolutions in Brodman's area, and others.

The Hobbit is not the only hominin with a small brain, contemporaneous with *Homo sapiens*, and exhibiting advanced behavior. A very new discovery is adding fuel to the fire. That is next.

A New Homo Species Found in 2015

Another astounding find occurred in South Africa in 2015. Professor Lee Berger, who we met in the previous chapter during our discussion of *Australopithecus sediba*, topped his own great discovery when he excavated a cave, called the Rising Star Cave, which was teeming with Homo lineage fossils. Mind you, Berger did not personally go down into the cave. Like most of us, he would not fit through the tight and tortuous tunnels leading to the fossil-containing chamber. Instead, he recruited six slim women having an anthropology education and an interest in excavation to go down into it for him. The National Geographic grant, which Berger obtained, lasted long enough for them to haul a treasure trove of fossils out of the chamber. If you would like to see the two-hour video of this fascinating story, you can find it on YouTube. Look for "The Dawn of Humanity," which is aired by NOVA.

Over fifteen hundred hominin bones were taken from the cave, and specialists from around the world were invited to participate in the analysis of the bones as they attempted to understand what they had found. The four skulls that were found show that the males had brain capacities of 560 cc, whereas the females averaged 365 cc. This is the range we find Australopiths within, but Berger decided that these hominins were definitely of the Homo lineage. In fact, he named the species *Homo naledi*. The shape of the skulls was human, as was the dental arch and teeth. The decision was confounded by the mixture of primitive and advanced traits exhibited by the specimens. They were bipedal walkers, but their pelvises were flared like the Australopiths.

One of the mysteries is how did all these bodies get into the cave. The only conclusion that seemed possible was that they were carried down by the *Homo naledi* themselves. However, that defied everything we thought we knew about human cultural development. Specifically, the *Homo naledi* hominins had a much smaller brain than we associate with ritualistic burials. That is a cultural behavior only associated with more recent humans, with brains the size of ours.

Until recently, we did not know when these hominins lived. Dating can be a problem with cave finds, and in this case, it was more so. Usually, there are other animal fossils found in caves containing hominins. If we can place those animals chronologically, we have a good idea how old the hominins were. Unfortunately, they didn't have that option for there were no animal fossils. Estimates based on skull morphology placed the species between 912,000 and 2,000,000 years old. In any case, they could not be older than 3,000,000 years because geologists tell us that the cave is only that old. But wait! Here is a news flash: Berger just announced in 2017 that he has narrowed the range to between 335,000 and 236,000 years. If that dating is accurate, these small comparatively primitive hominins may have existed concurrent with our own species. These recent finds certainly upset our previous beliefs about the drivers and constraints of human evolution. When these puzzles are finally resolved, paleoanthropology will be taught differently than it has been in the past.

Overview of Homo

Table 6.2 is a look at some of the various species, which fall under the genus Homo. Considering how recent discoveries like *Homo erectus georgicus, Homo florensiensis,* and *Homo naledi* have impacted and altered our view of the Homo lineage, it is a good bet that some future discovery will send us back to the drawing boards again. For example, our concept of Homo erectus has been broadened to think of this species name to include species as primitive as Homo habilis.

Table 6.2 Description of the Different Homo Species

Species Name	When Alive, mya	Avg. Brain Capacity, cc.	Comments
Homo sapiens	0.15 to now	1350	Us, the only remaining hominin
Homo florensiensis	0.19 to 0.05	400	Indonesian island, a dwarf homo erectus
Homo neanderthalensis	0.3 to 0.03	1500	Lived in Ice Age Europe and Middle East
Homo heidelbergensis	0.65 to 0.15	1200	Likely ancestor to Neanderthals and us
Homo antecessor	0.85 to 0.65	1100	Atapuerca, Spain, oldest European hominin
Homo erectus	1.8 to 0.3	900	Lived in Africa, Asia, and Europe
Homo georgicus	1.8	600	Dmanisi, Georgia, an early Homo erectus
Homo ergaster	1.9 to 1.4	690	An early Homo erectus, exclusive to Africa
Homo naledi	unknown	450-550	Rising Star Cave, South Africa
Homo habilis	2.3 to 1.4	640	Africa, few specimens, tool user.
Homo rudolfensis	2.5 to 1.9	700	Kenya, East Africa, single specimen

Homo erectus, the Runner

The Australopiths had skeletons that were not optimally designed for running. Their legs were too short, arms were too long, and they had a big gut. However, a bigger problem might have been their inability to cool their brains and bodies adequately during prolonged strenuous exertion such as running. *Homo erectus*, on the other hand, evolved into a proficient runner. In other words, *Homo erectus* had a body very much like ours, and we are excellent long-distance runners. Think about our marathon runners. They can run 26.2 miles in a little over two hours, but then it is only a sport for them. Whereas, it meant survival to *Homo erectus*. He had the ability to run prey animals down until they were exhausted and no longer able to escape.

Skeletal Changes

The earliest species of the Homo lineage resembled the Australopiths. They were about four feet tall, under one hundred pounds., with short legs, long arms, and a large gut. Evolving longer legs was one obvious improvement because longer legs mean longer strides. The landscape had changed so that there were few readily available trees to climb for refuge, so their arms got shorter. Their increasing reliance on meat for nutrition caused their gut to shrink. Predatory animals tend to have smaller guts than herbivores. The gut size difference is quite obvious if we compare the outward flaring ribcage of *Australopithecus* with that of the straight up and down ribcage of the entire Homo lineage since *Homo erectus.*

Adaptations to Cool the Body

As we discuss the evolutionary adaptations to cool the body, we will be also answering a few of the questions about the differences between humans and apes. One often asked is "Why do apes have fur, yet we do not?" Another might be "Why do humans have differently colored skin?" And finally, "Why do we have to drink water more frequently than other animals?" Scientists tell us that these remarkable changes in our bodies resulted from our need to become distance runners in a hot climate.

The scientist who has studied this question intensely and can explain it best is Nina Jablonski. She is both an American anthropologist and a paleobiologist**.** In her book *Skin, a Natural History*, Jablonski tells us that the *Homo erectus* clans of Africa apparently shed their fur in order to better cool their bodies and in particular better cool their brains. Their environment was a savannah-like landscape with extensive grasslands and few trees. Their daily activities involved strenuous walking and/or running for long periods, and those activities threatened to raise their temperature to a life-threatening level.

The main device that humans use for cooling is to sweat. Unlike apes, who sweat an oily fluid that coats their hairs, *Homo erectus* bodies evolved an abundant of eccrine glands over the surface of the skin, which produce a salty, watery fluid. Now remember that evaporation is a cooling process. This waterier sweat totally evaporates and thus more effectively cools the body. Think about our own bodies because we inherited this cooling system. It works very effectively except when the humidity is high, and then it is hard to get cooler. However, *Homo erectus* lived well over a million years ago and evolved these traits in Africa, which was a chronically dry environment. So, we see that they were adapting to a hot and dry climate over thousands of generations.

However, the fur had to go in order for this new cooling method to work. I imagine that some individuals had thinner body hair than others. The new eccrine cooling system worked better for them than for the more thickly haired individuals. Their survival ability was better, they had more progeny, and repeating this process over hundreds of generations reduced the hair increasingly and made us as hairless as we appear today. The story doesn't end there because as the body hair covering became very thin, new problems arose. Hair had protected the skin in the past, so now the bare skin introduced new adaptive problems to solve like sunburn, abrasion, and retaining warmth under cold conditions. One adaptation was for the skin to produce more melanin and darken the skin. Another was for the skin to become tougher and leachier. The fact that humans endured these adaptive hardships strongly suggests that cooling was highly important to our human ancestors, who lived in this environment. The bodies of modern humans are the way they are due to this adaptive phase of our biological history.

Adaptations to Cool the Brain

The person who probably knows the most about how the brain evolved during our evolutionary history is American Anthropologist Dean Falk. And she shares that knowledge with us in her book *Braindance*. She tells us that there were massive changes to the cooling systems for the brain. Although the actual brains of our ancient ancestors have decomposed, she is able to learn a great deal from studying endocasts of the fossil skulls. She concludes that the flow of blood was extensively rerouted to attain cooler brains during the *Australopithecus* and *Homo* lineage times. The end result was a superior cooling system for the brain compared to that of other mammals of the time. Man is not a very fast runner, but he is an excellent long-distance runner. Paleo anthropologists believe that *Homo erectus* and his progeny could run down game animals to their death. The prey animal's brains eventually overheated to the point where they were dazed, confused, and vulnerable to a deathblow to finish them off. The pursuers' brains were adequately cooled so that they retained clear minds and could deliver that lethal blow to the overheated animal.

Homo erectus, the Thinker

A comparison of table 5.1 in the previous chapter with table 6.2 of this chapter will lead to two general conclusions about the evolution of hominin brains. The first is that the long period of Australopith dominance was one of negligible change in brain size. The second is that the time from early *Homo* lineage all the way to us modern humans is one of spectacular brain growth. Even if we factor in the increase in body size during the Homo lineage, the mass-adjusted brain growth is still spectacular. Does brain size correlate with intelligence? I guess that the correct answer is not necessarily. We certainly have seen small-brained Homo species do things we didn't believe was possible. Yet, during the time *Homo erectus* lived, his or her brain size grew at a rate not seen before. The Hobbit and *Homo naledi* with their small brains tell us that we have more to learn. And yet, *Homo sapiens* dominate the earth today and bend it to our will. The genetic connection between large-brained *Homo erectus* and ourselves seems clear-cut.

The Archulean Tool Culture

Our prehistoric past might have remained forever hidden if were not for items from those times that resisted decomposition. Fossil bones are one type of preserved items, and stone tools are another type. What we see in the style and complexity of stone tools over our prehistory are valuable clues to our intelligence, culture, and innovativeness. The first stone tools were probably stones or rocks, which had fractured naturally to give a sharp edge suitable for cutting. These naturally sharp stones were not that easy to find, so it soon became more practical to smash one rock into another and thereby produce a sharp-edged rock on demand. Obsidian is glass made from volcanic action. It breaks with a conchoidal fracture pattern and can produce surgically sharp blades. Another material that behaves like that is flint (geologists call it chert), and most of the prehistoric stone tools found in Africa were made of flint.

The primitive stone tools found in the oldest strata are called "Oldowan" after the stone tools discovered by Louis and Mary Leakey in the Olduvai Gorge of Tanzania. A maximum age of 2.5 million years is commonly cited for Oldowan tools, but recent finds in East Africa are pushing that maximum age back to at least 3 million years. It may even become established that *Australopithecus afarensis* and other Australopiths made stone tools. It has long been thought that they were incapable of doing so. By 1.8 million years ago, a new stone tool-making culture called "Archulean" first appears in East Africa. These tools are quite symmetrical and may have a teardrop or oval appearance to them. The hand axes have been found in large quantities and are strongly associated with *Homo erectus*. Now, anyone examining these Archulean stone tools realizes that they are not an accidental product of nature but are the product of a planning mind and a skillful artisan. Something significant happened to the minds of these ancient toolmakers that tells us a leap in mental capability had occurred. We know from the fossil evidence, like that of 1.6-million-year-old Turkana Boy, that brain volume had increased compared to the earliest Homo lineage species. So, the two facts may be related. Now, here is another fact: the Archulean culture lasted a long, long time on the order of 1.7 million years. It was still practiced by some Homo erectus descendants one hundred thousand years ago. So how did the skills get passed on from generation to generation over such a long time? Some would argue that human speech must have been developed coincident with sophisticated toolmaking for that knowledge to continue.

An article in the April 2016 issue of *Scientific American* magazine by Dietrich Stout discusses the mental activities involved in creating a flint hand ax, similar to what *Homo erectus* and its descendant species *Homo heidelbergensis* made routinely in the past. He reports that it isn't as easy as you might think. He ran a school to train novices in the art of knapping. It took one hundred hours of instruction to prepare them to make stone tools such as the hand ax. The chief difficulty is learning to strike the work product with a hammerstone precisely and with the appropriate force to dislodge the desired chip. Stout measured brain activity in the participants and reported on the parts of the brain affected. This research work may provide support to the hypothesis that it was stone toolmaking, which was mainly instrumental in expanding human thinking ability.

The Taming of Fire

Homo erectus learned to tame and use fire. The archeological proof of this becomes more and more decisive in later periods of his existence. However, there is evidence in East Africa that could set the discovery of fire use back to 1.5 million years ago. By acquiring this capability, they gained a tremendous advantage in their struggle to survive. Fires at night kept predators at bay and provide warmth. This must have been vital during migrations to colder climates. Fire may have been used for hunting. Setting brush on fire can be used to drive prey animals off a cliff or into a killing ground. Cooking was learned sometime after fire was mastered, and that innovation had immense benefits to *Homo erectus* and all the Homo species that followed. Cooking makes foods more digestible, more nutritious, and easier and faster to chew. The human face has become flatter over time because jawbones, teeth, and skull structure have continually gotten smaller. Cooking also makes meats safer to eat by killing harmful bacteria and parasites.

The Hunting Mentality

The hunter-gatherer lifestyle may have begun earlier than previously thought according to a recent article (Sci. Am. Apr. 2014). It was previously believed humans obtained meat almost exclusively from scavenging in the beginning, and that actual killing of game was a much later human development. Cut marks on bones date back to 2.6 million years ago, but whether these animals were killed or scavenged was not known. Now evidence is being uncovered that sheds light on this type of question. For example, evidence of massive slaughter of wildebeests and other large animals had been previously found at a site in Tanzania, known as FLK Zinj. These animal fossils have been dated at 1.8 million years ago. The new element is that Professor Henry Bunt has discovered a way to distinguish between predator kills from human kills. Predators mainly select older, slower prey animals whereas man commonly takes animals in their prime. Therefore, a determination of the age distribution of prey animal bones can decide whether predator or human did the killing. The FLK Zinj site fossils were in their prime suggesting human hunting dates back to 1.8 million years ago. Another site called Kanjera South lies on the shores of Lake Victoria in western Kenya. This site is two million years old and contains numerous stone tools and butchered bones of young antelope. The evidence suggests that these animals were killed, not scavenged. It seems to me that these discoveries help us understand the existence of small-brained *Homo erectus* fossils in Dmanisi, which are dated at 1.8 million years ago. The ability to hunt makes it much easier for primitive human to migrate far from the homeland and survive. We also have to rethink the belief that hunting and meat eating began with the Homo lineage and not with Australopiths. The difference between these two groups is anything but certain at 2.5 million years ago and older.

Evidence abounds for us to say that hunting is associated with early *Homo erectus* and late Homo erectus and continued on with Heidelberg Man, then with the Neanderthals, and also with our own species. This two-million-year-old way of life persisted throughout the world and only was replaced when agriculture began its sweep beginning twelve thousand years ago. The ability to fashion stone weapons made hunting and butchering possible, but human thinking was the really important factor, which elevated humans to top predator status. Our ancestors had to outthink their prey. And that process becomes an arms race of sorts because prey animals react to new threats too.

We have seen that when man first encounters animals for the first time, they are approachable and easier to kill. But they learn over time to be wary of us, hide from us, and keep a safe distance from us. Once prey animals learned that man is dangerous, hunting became a lot more difficult especially if one's weapons must be used at arm's length. Throwing weapons can increase the hunter's yield. Heidelberg man was certain making them three hundred thousand years ago. We have evidence of spears shaped like a javelin from that long ago. However, weapon throwing may go back much earlier than that. The Archulean hand ax has always been a mystery because it has sharp edges all around its periphery. They must have been of value to Homo erectus because thousands of them have been found. Could it have been a throwing weapon? Another piece of evidence is the human body itself. We humans are powerful, accurate throwers, whereas apes come nowhere close to our ability. A comparison of skeletal features explains the difference. We have a more flexible waist, a longer thumb and stronger wrist, a less-twisted upper arm bone, and a sideways-facing shoulder. It seems that throwing has been important to our survival since we split off from our ape ancestry.

There had to be an element of coordinated teamwork to these hunts and that raises the question of when verbal communication originated. Hand signals and sign language have the advantage of being a silent communication. However, members of the hunting team must have been out of sight of each other at times, and a verbal signal must have developed. Perhaps, imitating a birdcall or chatter meant "Game in sight" or "Move toward me." It seems logical to me that at least the rudiments of a spoken language developed as hunting techniques evolved. Perhaps, the day's hunt was relived with pantomime and sounds over the evening campfire gathering.

Adaptation to Foreign Environments

As we have discussed already, *Homo erectus*, in one form or another, has been found on three continents. We think they originated in Africa because they must have descended from some Australopith species, and that ancestral species most assuredly originated in Africa. Now we find a small-brained version of Homo erectus existed in the European country of Georgia 1.8 million years ago. Moreover, these Dmanisi fossils are accompanied by stone tools, weapons, and butchered animal bones. That discovery throws a monkey wrench in the previous hypothesis where Homo erectus grows a big brain, develops an advanced tool kit, tames fire, and is then prepared to migrate to foreign lands. These puzzles are what makes paleoanthropology such fun.

New places harbor unknown threats and that makes them dangerous to visit. So, how was *Homo erectus* able to end up at places all over the world? Perhaps he was a very effective hunter and as such saw his tribe increase in number. Yet when hunting decreased the game supply, it became harder to feed the population. This kind of problem created impetus for some groups to split off and start their own clan. All these instances were motivations to seek new hunting grounds. So, we understand why Homo erectus was driven to enter new places, but the question is how did these individuals cope with the unforeseen problems that create. Higher general intelligence and cooperation would be needed to succeed. These drivers shaped the minds and culture of our ancestors of long ago.

Homo erectus, the Social Animal

Homo erectus is a species that underwent radical physical and mental transformations, and from those changes we can infer a great deal about their social life. Perhaps, the most astounding change was the doubling of brain volume and even tripling if we consider *Homo heidelbergensis* and *Homo sapiens* as extensions to the *Homo erectus* lineage. As their brain volume increased, so did the duration of time that their infants were helpless and totally dependent on their mothers. This is due to the fact that there is a maximum head size that can pass through the human birth canal. In order to attain a larger brain, most of the brain growth had to occur after the baby's birth. We know from our own experience that infants begin to walk only between one and two years old, and even young children must be constantly protected from danger for years. Young apes, by contrast, can survive on their own at a much younger age than this. So, we can conclude that there must have been a social structure to support mothers and young children. This would include a sharing of food and a sharing of childcare.

Today, long-term-committed male-female pairing is the social norm. We cannot say for sure when this custom began, but there is reason to think that it may have been practiced during the *Homo erectus* days. Chimpanzee females advertise when they are in heat by color change and swelling in their vulvas. Human females, by contrast, have hidden ovulation. Chimpanzee females mate with many males during their heat, with the result that none of the males know if they fathered the resultant baby chimp. However, human males would know if they were in a committed relationship with the pregnant female. This makes all the difference in providing an incentive to the human male to provide food and protection to his female and her offspring sired by him.

Moreover, there are other physical changes in humans, which are absent in apes, that tell us that romantic relationships were replacing the quick "Wham, bam, thank you Ma'am" kind of sex that apes do. This topic is covered in detail in chapter 8, so I won't repeat it here. Yet, the reader might think about the topic until then and consider how we are different in appearance and sexual behavior from the chimps and bonobos in the zoo. How did the increasing burden of dependent infants influence the male-female relationship?

Heidelberg Man

If there were a missing link between *Homo erectus* and modern man, that link is no longer missing. It is *Homo heidelbergensis* or more commonly, Heidelberg man. The name comes from a mandible found near Heidelberg, Germany, in 1907, but the richest source of fossil evidence comes from a cave in northern Spain. That cave is called Sima de los Huesos and is at Atapuerca. Since 1997, more than fifty-five hundred skeletal remains have been excavated at this site. These remains have been dated at 350,000 years old and represent twenty-eight individuals of the species *Homo heidelbergensis*. The cranium in figure 6.3 is from Sima de los Huesos. This individual, who has a very human appearing cranium, obviously suffered impacts that proved fatal to him.

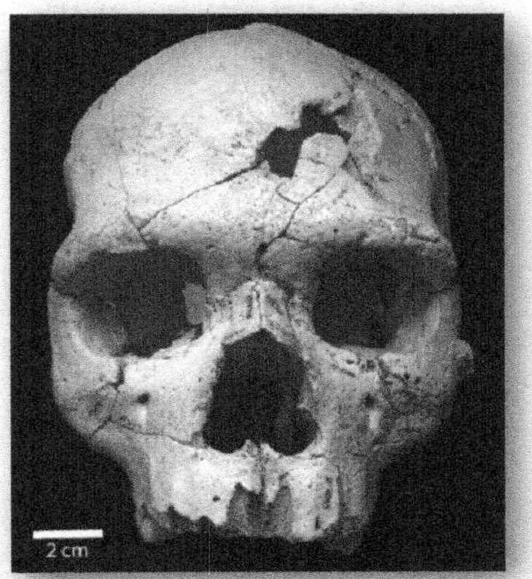

Figure 6.3 Heidelberg Man

By Sala, Nohemi, Juan Luis Arsuaga, Ana Pantoja-Pérez, Adrián Pablos, Ignacio Martínez, Rolf M. Quam, Asier Gómez-Olivencia, José María Bermúdez de Castro, and Eudald Carbonell. "Lethal Interpersonal Violence in the Middle Pleistocene." *PLoS One* 10, no. 5 (2015): e0126589. http://journals.plos.org/plosone/article?id=10.1371/journal.pone.0126589, CC BY 4.0, https://commons.wikimedia.org/w/index.php?curid=49818575.

Heidelberg man originated in Africa perhaps 700,000 years ago. Fossils have been found in Ethiopia, Namibia, and South Africa. Groups of these people migrated into Europe and Asia some 400,000 to 300,000 years ago. The brain of Heidelberg man, with a volume of twelve hundred cubic centimeters, was nearly as big as the brain of modern man. Indeed, we modern humans most likely descended from the African members of this species. However, we are not the only hominin species claiming him as an ancestor. The Neanderthals of Europe and the Middle East are descendants, who became especially adapted to surviving under Ice Age conditions. There exists another species, the Denisovans, which adapted to conditions in Asia. Genetic evidence suggests that the split between Neanderthal and Denisovans occurred over 430,000 years ago.

Here are some of the things that we know about Heidelberg man: The males averaged five feet nine inches in height and the females, five feet two inches. As for weight, the males averaged 136 lb. and the females 112 lb. These people were hunter-gatherers and capable of making spears with hafted spear points. There is also evidence that they buried their dead.

Neanderthal Man

Neanderthal fossils were being discovered in Europe back in the mid-nineteenth century, when Charles Darwin was writing his impactful book *On the Origin of Species*. Since then, fossils of this sister species have been found throughout Europe and the Middle East. For years, it was thought that we might have descended from them. Now we have evidence that they evolved on a parallel path to us. The official species name for them is *Homo neanderthalensis*, but it is easier to just call them Neanderthals. I think of them as the most interesting human species other than us. They lived at the same time as our ancestors and interacted with them before they went extinct some thirty thousand to forty thousand years ago. They even mated with our ancestors. We carry about 2 percent of Neanderthal DNA in our genomes yet today.

So, what were they like? The term "Neanderthals" carried a negative image for decades based on a fallacious perception of them. In fact, many of us still believe the word "Neanderthal" denotes a stoop shouldered, primitive, brute. The truth is that they were a lot more like us than you might think. They had a bigger brain on average than we *Homo sapiens* do. However, their skull shape was somewhat different from ours, so there are differences in the brain matter. For example, our frontal lobes are bigger than theirs. If a Neanderthal was dressed up in a suit and tie (or skirt and blouse in the case of a female), he or she might pass as a little strange looking but more likely would get some serious stares. Their noses would probably attract attention first. Neanderthal noses and breathing system are unique in the entire Homo lineage. They were quite large and seem to be specialized to retain warm, moist air. This ability was probably important to survival in an Ice Age world. Like *Homo erectus* and *Homo heidelbergensis*, they had prominent eyebrow ridges. This would be another noticeable trait. They lacked the prominent chin of our species. They had stocky bodies with thicker bones. They were probably much stronger than we are.

We know a little about the way they lived. They were ambush hunters, who speared large animals at close range. Their skeletal parts display the same kind of broken and mended bones that we see in rodeo riders of today. They had a distinct stone-tool culture, called "Mousterian," which seems to be a derivation of the Archulean culture associated with *Homo erectus*. In other words, it was an advanced toolmaking technique. On the other hand, they were not very innovative; this tool culture stayed the same for a hundred thousand years or more. Our more recent ancestors, by contrast, were constantly adapting their tool types to the varying circumstances. The Neanderthals had control of fire and may have cooked their meat. They cared for their wounded. Crippled individuals lived for decades after their wounds had occurred. They must have been cared for by others.

Denisovans

We mentioned that another species is believed to have evolved from Heidelberg man's migrations to Asia. We said they were the Denisovans. Surely, there must be a wealth of fossil evidence to help us describe them to you. The last that I heard, we only had a finger joint and a tooth. You may be wondering right now, how we could possibly claim to have discovered a new species on such little evidence. That story must wait until the next chapter.

Seven

Nucleotides, Migration, and Hybrid Genes

The Chapter in a Nutshell

Writing only reaches back five thousand years, and although archaeological finds reach back millions of years, such finds are rare. Clearly, we need yet another tool if we are going to resurrect our primitive past. Luckily, modern science has given us a powerful one. It is the DNA molecule, which lives in the nucleus of the cells in our bodies. This gigantic molecule has acquired thousands of so-called markers at various times in our ancestral past, which when analyzed along with the distinct markers of other humans can reveal important facts about our human history.

"Mitochondrial Eve" was the first accomplishment of DNA testing. Mitochondrial Eve, or mtEve for short, is defined to be the common female ancestor to all living humans. In other words, she is directly related to every person on the planet. The question was "How long ago did she live and where did she live?" Now these researchers focused on the much smaller DNA molecule we also possess. That is the mitochondrial DNA residing outside the cell's nucleus. Samples of mtDNA were taken from women in locations around the world, and from the markers in them, the age of that last common female ancestor was determined. She lived between 140,000 and two hundred thousand years ago in Africa. This use of mtDNA to reach into our distant past became a landmark discovery, and mtEve became famous as the story spread.

Y-chromosomal Adam became mtEve's male counterpart, and whereas mitochondria are a female genetic link, the Y chromosome is a male genetic link. It was a step up in DNA technology. Although much larger than mtDNA, the Y chromosome is the smallest of our forty-six chromosomes. Where the mtEve effort used only women in the study, this effort used only men (women lack the Y chromosome). Sampling, testing, and analysis show that Y-Adam is equally as old as mitochondrial Eve, although it is unlikely that they lived at the same time or ever met each other.

Tracking Out-of-Africa migrations of *Homo sapiens* was the next use of the DNA tool. In separate efforts, first mtDNA and later Y-chromosomal DNA analysis techniques were employed to track the migratory history of our species as it left Africa. The detailed routes of the "Out-of-Africa" migrations, which began over fifty thousand years ago, and eventually took us to all the habitable continents of the world, represent one of the major accomplishments of science.

Sequencing the human genome was a goal of the Clinton administration, but Svante Paabo had a bigger goal in mind. He devoted his life to restoring and analyzing the DNA of long-dead animals and humans. He had phenomenal success in this regard but not without having to overcome horrendous obstacles along that path. In the end, he was able to sequence the nuclear DNA of several Neanderthal fossils and establish that it is slightly different from our DNA. He found about 2 percent Neanderthal DNA in our genomes. Finally, based on the DNA in a tiny finger bone from a cave in Siberia, he discovered a new human species dubbed the Denisovan Man. Denisovan DNA shows up in South Asians particularly.

Understanding the Power of DNA

Bountiful Information

Due to recent fossil finds, we know that our species, *Homo sapiens*, goes back at least three hundred thousand years. However, by employing the power of DNA tracking, we also know that it originated in Africa and subsequently migrated out of Africa and occupied all the continents of the Earth except Antarctica. Even today no one lives on that frigid continent except for research expeditionary staff. Moreover, we know from DNA tracking about when each continent became occupied by our species even though these migrations happened long before writing had been invented. As for physical evidence of the migrations, such as fossils and stone tools, that evidence is quite sparse. So you might ask, "How then could we possibly know such detailed things about our past? It seems like those things are beyond knowing." The answer is that the information has long existed in our collective human bodies. We only learned how to extract and interpret it in recent times. How we did it is a story of great scientific accomplishment, which too few people know of, or appreciate. Perhaps one of the reasons for its obscurity is that it gets a bit technical. However, I think that understanding this new power of DNA tracking is so important to all of us, that its story needs to be told. What we have learned about DNA is a story of great human achievement in solving the elusive mystery and the actual mechanism of reproduction and inheritance. The accomplishment is a thing of beauty.

The Secret of Life

We humans have some one hundred trillion cells in our bodies, and within the nucleus of those cells is a most remarkable material, namely, DNA. For those who care, the term "DNA" is short for the chemical name, deoxyribonucleic acid. However, what is important is knowing what DNA does. Among other functions, it controls our biological inheritance, and it builds, rebuilds, and maintains our bodies. We know a great deal about DNA, but that knowledge didn't come easily or fast. It took several generations of scientists, each building on the investigative work of their predecessors, to get us where we are today. The crowning event of this collective enterprise was the modeling of the three-dimensional DNA molecule. You can read about this intriguing tale in *The Double Helix* written by James D. Watson. It is a David-and-Goliath kind of story, where two unknown scientists, Crick and Watson, in Briton were competing against the world-renowned American chemist, Linius Pauling. Moreover, while Pauling had great resources at his disposal, Crick and Watson worked on the problem without full management approval. Despite the disadvantage, the British team were successful. They built a physical model of the DNA molecule, and it was a double helix. Two strands of polymeric material spiraled around each other. When they realized what an important pivotal step in solving the problem of biological inheritance they had done, they said they felt like they had discovered the very secret of life. Such is the beauty and simplicity of how the molecule can replicate itself and perform its life-controlling functions.

Essential Facts about DNA

The Double Helix

DNA can be described in different ways, but each description adds to our understanding of this miraculous substance. For one, the DNA of any individual is unique. Television watchers already know that DNA is a better identifier of an individual than are fingerprints. They also know that you can get a DNA sample from a cheek swap, a drinking glass, or a hairbrush. However, there is a lot more to know. Consider its physical shape. Structurally, DNA is a double-stranded helix (see figure below).

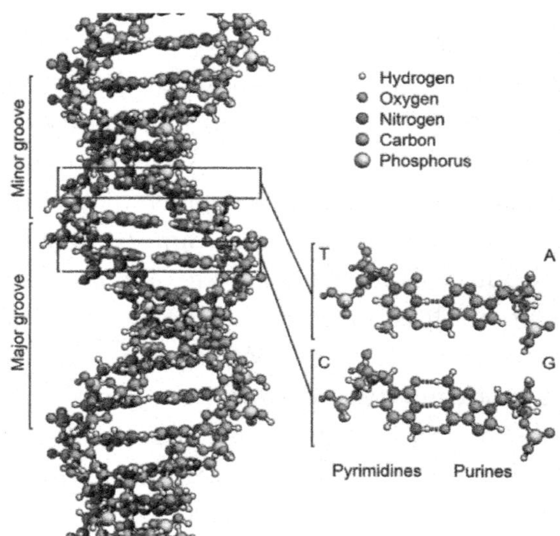

Figure 7.1 The DNA Double Helix
Courtesy of Wikimedia Commons

That is, the two molecular strands spiral around each other to form what might appear to be a spiral stepladder. We should also consider DNA's enormous length. DNA is a very, very long molecule having repeating units called nucleotides. For either one of the two spiraling strands, the number of nucleotides in that strand is about 3.2 billion. If you count both strands, the total number of nucleotides is 6.4 billion. Of course, words like "enormous length" are a matter of scale. We have trillions of cells in our body, so any of those cells and the DNA within in them are microscopic relative to our human size. However, if we could shrink ourselves to the size of a carbon atom, the DNA strand would seem incredibly large.

The Code of Life

Here is another fact about DNA, and this fact is stunning; each nucleotide has its own unique code letter, and it can be one of the four types: *A*, *T*, *G*, and *C*. Together with the other nucleotides in the string, coded messages exist that are essential to life! Remember how, in previous chapters, we talked about the important of genes to evolution. To the nineteenth-century monk, Gregor Mendel, the gene was the discrete unit of inheritance. He knew it existed based on his studies of inheritance. Now we look at a gene from a structural standpoint. Now we see genes as an instruction code made up of a defined sequence of nucleotides, and that image is the essence of genes as seen from a biochemist's viewpoint. Genes are the instruction codes for the cell to produce specific proteins. The proteins carry out the life functions.

DNA seems to have magical powers when we think of all that it does. However, scientists have demystified most of its functions. It turns out that only a fraction of the DNA molecule is doing actual work, while the remainder merely goes along for the ride. That nonfunctional part is called "junk DNA," whereas the functional part consists of genes and switches. Those genes and switches affect inheritance and all other aspects of life. How is this significant? Well for one thing, these coded messages (or genes) are what cause a baby pup to look like its canine parents, a baby kitten to look like its feline parents, and a human baby to look like its human parents. These coded messages can also rebuild your dying cells, fight infection, take you through adolescence, and every other function of life. You can see these code letter designations in the figure of the DNA molecule above.

Cell Division

In building their model of the DNA molecule, Crick and Watson had to account for DNA's ability to duplicate itself during cell division. After all, we begin life as a single cell composed of our mother's egg and our father's sperm. It is through cell division that we grow. The significance of their two-stranded model can now be better understood. DNA can duplicate itself prior to cell division by a process illustrated in the following figure. The illustration begins with the DNA helix separating into individual strands. The second circle shows the cell manufacturing a new strand on each of the individual strands. The remaining circles show the process of mitosis or formation of two new cells.

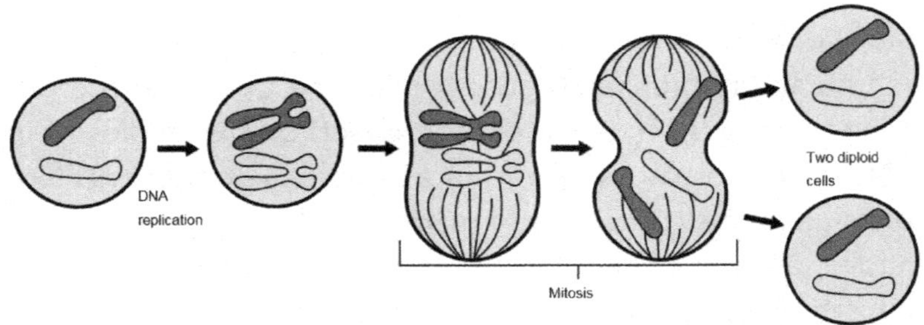

Figure 7.2 DNA Replication

By Mysid—Vectorized in CorelDraw by Mysid from http://www.ncbi.nlm.nih.gov/About/primer/genetics_cell.html., Public Domain, https://commons.wikimedia.org/w/index.php?curid=1414818.

It is the events in the second circle that Crick and Watson attempted to understand. How is it that the DNA molecule is able to perfectly duplicate itself? They learned that nature follows some simple rules in building those second strands. Imagine yourself starting with the first nucleotide on the strand and looking for its code designation: If the nucleotide is a "T," then pair it with a nucleotide designated with an "A"; if it is an "A," then pair it with a "T." If it is a "G," then pair it with a "C," and finally, if it is a "C," pair it with a "G" on the duplicate strand you are building. Now advance to the next nucleotide and repeat the process.

Perhaps, a more familiar item can help illustrate this concept better. Our item will be a beaded necklace rather than nucleotides. Now imagine four bowls full of colored round beads. One bowl has red beads, one has green beads, one has blue beads, and one has yellow beads. Now someone has already arranged colored beads on a wire, and the arrangement appears to be random. For example, a segment might be "RRYGBBGYYYR." Your job is to arrange beads on a second wire, while following these rules: If you see a red bead on wire #1, place a green bead on wire #2. If you see a blue bead, place a yellow bead. If you see a green bead, place a red bead, and if you see a yellow bead, place a blue bead. Now let's say that green and red beads have a special magnetic attraction to each other, and yellow and green beads have an attraction for each other. As you add beads to wire #2, you will notice that the new bead, if correctly chosen, will attach itself to complementary bead on wire #1. However, if you select the wrong color, there is no attraction at all. This is something like what DNA replication is like but on a molecular level.

Polymorphisms

We all inherit our genes in pairs, one from our father and one from our mother. Those two genes might be identical, or they might be slightly different. So, the traits we inherit depend on whether a given gene is recessive or dominant. Of course, trait inheritance is not always as simple as Mendel's results suggested. Most traits are not controlled by a single gene but by a number of genes. When multiple genes control a trait, it is then possible to get a trait that is intermediate between those of the parents. Let me give you a personal example: My first wife, who was the mother of my four children, had red hair and green eyes. I had blond hair and blue eyes (my hair is gray now, but my eyes are still blue). All four of our children had reddish-blond hair to various degrees. All four children had blue eyes, although my youngest had bluish-green eyes. His two eyes seemed to be identically colored at a glance, but if you looked closely, one eye was slightly greener than the other eye. Genes control inheritable traits, and the science of inheritance is being discovered every day in the laboratories around the world.

If you do any reading on genetics these days, you will invariably run into the term "polymorphism." It has to do with different morphs or forms of a biological entity. Remember that there is a relationship between the trait and the genes, which produced it. The outward appearance (or "phenotype") is determined by the genetic component (or "genotype"). Now, since we inherit our genes in pairs (one from dad and one from mom), if those paired genes are different from each other that is a polymorphism.

The classical example of a powerful polymorphism is the sickle-cell gene. If you live in an area where malaria is prevalent, the gene might be beneficial to you. If you receive the sickle-cell gene from only one parent, you will have better immunity to malaria than if you had normal genes from both parents. However, if you receive the sickle-cell gene from both parents, you get sickle-cell anemia and may die. It sounds unjust in a way. A roll of the dice determines whether you have no protection to malaria, great protection, or a fatal disease at birth. Unjust or not, the polymorphism is self-sustaining as long as malaria is a threat. However, if the region in which you live has no malaria threat, the sickle-cell gene will eventually disappear in future generations.

Protein Manufacture in the Cells

Okay, we get the idea that genes control inheritable traits, but we still haven't described how a gene exerts its influence. The answer lies in the domain of biochemistry and from that viewpoint, a gene is essentially a coded instruction list for making specialized proteins. In other words, a gene contains a particular sequence of coded nucleotides, which the cell's machinery interprets and uses to build a very unique protein. All living things have genes, and humans have about twenty-one thousand of them all contained within their genome. So, now we know that those genes are actually instructions for building a particular protein molecule in the cells. But so what? Why are proteins so important? Proteins are vitally important because they are required for the structure, function, and regulation of the body's tissues and organs. They serve as antibodies, enzymes, hormones, body structure, and transporting agents. In other words, genes do their magic by defining, building, and deploying very unique protein molecules, and those proteins, in turn, build, maintain, and repair our bodies. Some genes, called regulatory genes, do their magic by activating or deactivating other genes. They accomplish this feat by triggering the gene's switches. This also is a biochemical process.

We don't need to get into the fine details of protein synthesis. However, there are a few fundamental facts that are useful to know: Proteins are long molecules assembled from a particular sequence of different amino acids. Think of amino acids as the building blocks of living tissue. It turns out that there are twenty amino acids that are used by all living things, and that fact should give you a sense that all living things are closely related. So, the most important function of the gene is to specify the precise order in which these amino acids are lined up and linked together. In order to uniquely identify one of those twenty amino acids, the code is read three nucleotides at a time. These three nucleotide information packets are called "codons." For example, the amino acid alanine is identified by the codon "GCC," and the amino acid glutamine is identified by the codon "CAA." The process of finding the correct amino acid and getting it to the assembly site is done by molecular entities called "transfer RNA." When the correct number of amino acids in the correct order has been lined up, they are triggered to react with each other. In other words, they link together and form what is called a polypeptide. Protein can be composed of one of several polypeptides. Much of the power of the completed protein comes from its shape, and the right assembly of the right amino acids determines that shape. Protein folding is a topic of great interest to biochemists these days because we humans are using these cellular processes to manufacture medical drugs. That is right, biochemists have learned how to put that cellular machinery to work for us in making the proteins of our choice. For example, human insulin is one of the success stories of the biotech industry. Wikipedia has a wealth of information on the details of these processes if you would like to learn more.

The DNA Record of Our Migrations

Mutations Are Historical Markers

We almost know enough about DNA now to understand how we are able to resurrect our ancient migratory history. I am talking here about human migrations that occurred long before writing was invented. However, there is one last thing that we need to know about and that is mutations. It is the accumulated mutations in our DNA that serve as a historical record of our past. The mutations acquired by our most ancient ancestors have been passed on to us over the generations. So, what is a mutation? A mutation is a change in the DNA letter sequence. That is, a change from what the sequence always had been. The change might be a deletion, an addition, or a letter swap. Here are some examples. Assume the original sequence was "AAGCCTCG."

1. Deletion: new sequence becomes "AAGCTCG"
2. Addition: new sequence becomes "AAGCCATCG"
3. Letter swap: new sequence becomes "AAGCTTCG"

Mutations are very rare and normally disappear when the affected individual dies. However, if the mutation is in the germ line (sperm or egg), then future generations may carry it. Mutations may occur in our junk DNA, and the effect of the mutation is neutral. However, if the mutation occurs within a gene, the result might even have fatal results because a vital protein can no longer be correctly made. If the mutation happened to a gene in sperm or egg, the offspring may be affected from mildly to severely. We can categorize mutations as being harmful, beneficial, or neutral. Harmful mutations tend to be eventually purged from the populations due to natural selection. Beneficial mutations are rare, but they account for the optimization of a species as it adapts to a particular environment. Some mutations to our genes can be neutral, but all mutations to our junk DNA are neutral. Due to the fact that most of our DNA is junk DNA, these mutations provide most of the markers, used to interpret our historical past. Okay, now we know enough about DNA to follow the work of scientists using it to unravel our hidden past.

Applying DNA Technology to Human Evolution

Blood Types

Scientists, attempting to understand the migrational past of humans, were limited in what they could learn by the state-of-the-art tools available at the time. Even before we knew about the power of DNA sequencing, we had tools with at least some discriminatory power. Take blood types, for example. Humans have different blood types as everyone knows, and that by the way is an example of a polymorphism. In the days before DNA science, blood types were one of the few tools available to trace human prehistory. The idea is to compare the frequency of certain blood types in different geographical locations and to see what can be inferred from that information. For example, studies tell us that Native Americans have a very high frequency of Type O. The B blood types are the rarest of the ABO blood types and show up heavily in central Asia and somewhat in Africa. The A blood types show up most heavily in Europe, Australia, and Arctic populations.

Mitochondrial DNA

Blood types told us something, but a better tool became available in the 1980s, which yielded a lot more information. That new tool was the genetic record found in mitochondrial DNA (mtDNA), a type of DNA found outside the nucleus of other cells. Mitochondria are vital for converting food energy into physical energy. One advantage to using mtDNA is that it is consistent between generations because it is inherited solely from our mothers, whereas nuclear DNA is a mixture of the DNA inherited from both parents. Moreover, mtDNA mutates faster than nuclear DNA and thereby provides more markers. However, in order to find those mutational markers, we first have to read the genetic code of the DNA molecule, and reading the code was a slow tedious process in the 1980s before fast sequencing machines were invented. While it was then impractical in the 1980s to sequence the massive nuclear DNA molecule, sequencing the much smaller mtDNA was feasible. The mtDNA, being only 16,500 nucleotides in length, is actually tiny when compared with nuclear DNA that is 3.2 billion nucleotides in length.

So analysis of the data from mtDNA is much simpler than is the analysis of the data from nuclear DNA. Other advantages are that mtDNA is more easily acquired, and that it has more frequent markers to differentiate time periods. The first application of genetic testing using mtDNA led to determining the age and geographical origin of "Mitochondrial Eve." This famous research work is known as the Wilson, Cann, and Stoneking study. However, fictitious Mitochondrial Eve may be more famous. So, what is she? Mitochondrial Eve, or mtEve for short, is defined to be the most recent, common female ancestor to all living humans. In other words, she is directly related to every living person on the planet as a kind of great, great...great-grandmother. The particular question that the researchers were trying to answer was "How long ago did she live and where did she live?" Samples of mtDNA were taken from the placentas of numerous pregnant women in locations around the world. The mtDNA was sequenced to identify the markers in them, and using computers to establish the relative age of the markers, the age of that last common female ancestor was determined. She lived between 140,000 to 200.000 years ago, and she lived in Africa. This use of DNA to reach into our distant past became a landmark discovery, and mtEve became famous as the story spread. Recently, fossil evidence has pushed our species age back to 300,000 years and that is consistent with their findings. MtEve is not the Eve in Genesis. She was not the first woman. MtEve had a mother, grandmother, and female relatives far back in time carrying that same mtDNA.

Y-Chromosomal DNA

Y-chromosomal Adam became mtEve's male counterpart, and where mitochondria serve as a female genetic link, the Y chromosome serves a male genetic link. What is so special about the Y chromosome? The Y chromosome contains genes that make an embryo male. Both males and females inherit an X chromosome from their mothers, but the father's sperm determines the sex of the baby. If the sperm carries an X chromosome, the baby will be girl, and if it carries a Y chromosome, the baby will be a boy.

So, seeking the age of Y-chromosomal Adam is the male equivalent of seeking the age of mitochondrial Eve using mtDNA. Now chromosomes come in pairs, and both men and women have one X chromosome. If that second chromosome is an X, you are female, and if it is a Y, you are a male. By using the Y chromosome, we trace only male inheritance, and that makes the job of data analysis to determine Y-Adam's age feasible. The state of the art in DNA sequencing was improving, and although the Y chromosome is part of our nuclear DNA, it is a small part of it. On the other hand, it is much larger than mtDNA and consequently provides more information. A DNA strand on the Y chromosome is about fifty-eight million nucleotides in length. As you might surmise, there is much more information potentially available on the Y chromosome than on the much smaller mtDNA molecule due to the vast size difference. However, it was the invention of fast sequencing machines that made sequencing the Y chromosome feasible. These machines were invented to support the Human Genome Project that began in 1990 and was completed in 2003. Sampling, testing, and analysis show that Y-Adam was two hundred thousand to three hundred thousand years old. That range is from a 2015 calculation. So, he is of a similar age as Mt-Eve, although it is extremely unlikely that they lived at the same time or ever met each other.

Complete Genome Sequencing

The final advance was to develop the fast sequencing equipment, computational skill and equipment, and interpretive ability to compare entire genomes. We are into that phase of the new science in the early twenty-first century. In the 1990s, we sequenced the human genome; now we sequence thousands of human genomes from populations around the world and provide have these data available to researchers around the world. It is an amazing story of human achievement.

The Hominoid Trichotomy Revisited

In chapter 5, we introduced the once-unresolvable problem of whether we descended from African gorillas or African chimpanzees. We now look at the problem again in the light of our newer understanding of DNA. My source here is Eugene Harris's book *Ancestor's in Our Genome*, pages twenty-five to thirty-five. He presented the evidence better than I am able to do here, so if possible, read his book for the most complete and detailed explanation.

The Species-Tree Question

The three living species humans, chimps, and gorillas all descended from a common ape ancestor according to DNA comparisons and what we know from mammalian evolution. The question is which species split off first. If we can determine this, we will know to which species we are genetically closer. This can be expressed as a tree, and there are three possible outcomes:

> Tree 1. Gorilla branching off first; chimp and human branching off last.
> Tree 2. Chimp branching off first; gorilla and human branching off last.
> Tree 3. Human branching off first; chimp and gorilla branching off last.

Gene-Tree Experiments

Researchers tried to resolve the species-tree problem by running gene-tree studies of the three species. Humans, chimps, and gorillas in most instances use the same genes for the purposes. Such genes are called homologous genes. Since our common ancestors lived millions of years ago, there has been ample time for mutations to occur in those genes. If any mutations are common to all three species, they must have occurred prior to any of the three species splitting off. If any mutation is common to only two of the three species, then the species without that mutation probably split off first. The more times we see this pattern in other mutations, the more confidence we have in our conclusion.

Numerous different homologous genes were sequenced and compared, but the results were inconclusive. Some genes confirmed Tree 1, others Tree 2, and still others Tree 3. The problem seemed unsolvable and thus the name "The Hominoid Trichotomy."

Final Resolution

After the Human Genome Project was completed, it was possible to sequence entire genomes, and scientists used the new capability on unsolved problems. Thus, the entire genomes of human, chimp, and gorilla were compared, and the dominant trend confirms species Tree 1. The gorilla split off first, and we are more closely related to the chimpanzee. The actual results are that two-thirds of the genome confirmed Tree 1, the remaining one-third of the genome confirmed either Tree 2 or Tree 3.

The Out-of-Africa Migration

Spencer Wells

Spencer Wells is a geneticist, anthropologist, and entrepreneur. He has been a dynamic force in using genetic information to determine the unwritten history of our species. With funding from National Geographic, he embarked on the Geographic Project from 2005 to 2015. He applied the power of Y-chromosomal techniques to unraveling the story of our "Out-of-Africa" migrations and chronicled them in his book *The Journey of Man*. That journey began in northeast Africa at a time when there were only a few thousand of our Homo sapiens species in existence. However, these very innovative and hardy people proliferated and filled the planet with their kind. Wells identified particular haplogroups at various spots in the journey using the relevant Y-chromosomal markers. What is a haplogroup? Remember we talked about inheritable mutations and how they might define a group of people having that marker? Those people are a haplogroup. Well, the original haplogroup for the "out-of-Africa" migration comprised a people in northeast Africa (Haplogroup M168). Wells referred to the identifiable groups by a geographical name and by the haplogroup identification name. He also gave an estimated age for the origin of each haplogroup. Table 7.1 summarizes the "Out-of-Africa" migration. Here is a brief summary of the migration:

Table 7.1 Y Chromosomal Haplogroups in out of Africa Migrations

Haplogroup	Name	Comments	Previous	Next
M168	Eurasian Adam	31 K to 79 K years ago Northeast Africa		M130, M89
M130	Coastal Route Clan	Australia, Mongolia, Siberia, America, and Pacific Islanders	M168	
M89	Middle East Clan	30K to 50K years ago The Levant	M168	M9
M9	Eurasian Clan	40 K years ago Widely dispersed group	M89	M20, M45
M20	Indian Clan	30 K years ago India	M9	
M45	Central Asian Clan	35 K years ago	M9	M175, M173
M175	East Asian Clan	Korea, etc.	M45	
M173	European Clan	30 K years ago 70% of S. Englishmen	M45	
M242	Siberian Clan	20 K years ago	M45	M3
M3	American Clan	90% of S. Americans 50% of N. Americans	M242	

Middle East

The Middle East Clan (M89) was descended from Eurasian Adam (M168), some fifty thousand to thirty thousand years ago and resided in the Levant, an area east of the Mediterranean Sea. The Eurasian Clan (M9) descended from the Middle East Clan (M89) about forty thousand years ago.

Asia

The Coastal Route Clan (M130) was also descended from Eurasian Adam (M168). However, the group that settled in India, The Indian Clan (M20), descended from the M9 group, which descended from the

M89 group. India was settled about thirty thousand years ago. Another group descended from M9, namely the Central Asian Clan (M45) settled in China about thirty-five thousand years ago. The M45 group fostered two descendent groups, namely the East Asian Clan (M175) and the European Clan (M173) about thirty thousand years ago. The East Asian Clan settled in Korea and surrounding area. Finally, there is the Siberian Clan (M3), which descended from the Central Asian Clan (M45) about twenty thousand years ago.

Europe

The migration did not lead to Europe by a direct route as might be expected. Instead, the migration led to central Asia first, where part of the M45 clan turned west toward Europe around thirty thousand years ago and acquired markers that made them the European Clan (M173). We see this marker in 70 percent of today's Englishmen. It was the European clan that also became known as the Cro-Magnon and left bountiful artifact evidence of becoming human. This includes cave art, statuettes, beaded jewelry, bone flutes, and advanced tool- and weapon making. They also had encounters with the Neanderthals, a separate subspecies.

Americas

The American Clan (M3) descended from the Siberian Clan (M242). It is believed that they (M3) migrated across the Bering Strait around fifteen thousand years ago when it formed a land bridge between the continents. Ninety percent of South Americans are M3s, whereas only 50 percent of North Americans are. A second wave of different immigrants apparently diluted the M3s.

Pacific Islands

Coastal Route Clan (M130) descended from M168. It was the Coastal Clan people who populated the Pacific Islands in their outrigger canoes.

Out-of-Africa Summary

That brief summary certainly doesn't do justice to the beautifully told story of the migration in *The Journey of Man*. What I hope it does accomplish is illustrate to you the power of DNA technology in extracting information about our prehistoric past. Consider this brief summary a sampler to whet your appetite for the full story.

Paleogenetics—Sequencing Ancient DNA

Science Fiction

Michael Crichton (1942–2008) was one of my favorite authors and a man of extraordinary knowledge and imagination. He wrote a science-fiction book, which became a movie called *Jurassic Park*. I suspect that most of you have seen it and its sequels. In the original story, the DNA of dinosaurs is retrieved from mosquito blood. The blood exists in the bodies of mosquitos embedded in amber formed millions of years ago. The dinosaur DNA is injected into the egg of a modern reptile, and the long-extinct dinosaur returns as the egg hatches. That scenario is imaginative but not impossible. Each of the steps in the process is realistic. The technical weakness is that DNA decomposes with time and has surely disappeared after sixty-five million years.

Extraordinary Researcher

While that story was science fiction and a fantastic leap into a possible future, there is a real scientist, who has done what seemed impossible too. His name is Savante Paabo, and his life story is one of extraordinary accomplishment. Savante Paabo was born in 1955 in Stockholm, Sweden. He earned his PhD in biology at Uppsala University in 1986. Since 1997, he has been director of the department of genetics at the Max Planck institute for Evolutionary Anthropology in Leipzig, Germany. However, these statistics tell us little about his historic place in science. You see, Paabo is the father of a new science called paleogenetics, that of restoring and sequencing the DNA of extinct species.

Like the other researchers mentioned above, Paabo used shorter DNA samples at first and progressed to larger segments of DNA. After all, the way to eventually solve a large complex problem is to start out by solving smaller, simpler problems. That is exactly what Paabo and his teammates did. They did their early work on mtDNA rather than nuclear DNA. The mtDNA is only 16,500 nucleotide pairs long, whereas nuclear DNA is 3.2 billion nucleotide pairs long. Until ultrafast sequencing machines were developed, entire genomes could not be sequenced.

Obstacles to Success

Sequencing the DNA of ancient humans and animals was anything but easy, and he endured many failed experiments. Yet, Paabo, against incredible odds and with a "never say die" spirit, has blazed the trail of this new field of science. The obstacles to success fall into two main categories; decomposition and contamination.

Decomposition: The first problem is that DNA decomposes over time. Paabo attacked this problem by cutting his teeth on recently dead specimens like mummies and the famous Ice Man of the Alps. Later, he looked for fossils in cold climates where decomposition has been retarded. He also discovered chemical tricks in reversing the decomposition. Decomposition makes analysis harder because the DNA is broken into numerous small segments due to oxidation and microbial action over the years. It is a major undertaking to figure out how to reassemble the pieces. The problem is a bit like constructing a five-thousand-piece jigsaw puzzle but even more difficult than that.

Contamination: The second problem is that microorganisms often invade the ancient DNA thereby contaminating it. The problem of contamination was attacked by using painstaking precautions to avoid contamination. Particularly, it is vital to prevent contamination from the people handling the samples. He employed state-of-the-art contamination-control procedures like laminar-flow clean room. The personnel were specially trained, and testing for contamination was thorough.

Accomplishments

Paabo and his team were successful in sequencing the DNA of Neanderthal fossils. Their first attempt was using mtDNA, and the data was useful in estimating the age of our common ancestor. However, as the human genome project was getting completed, he saw the greater importance of sequencing the nuclear DNA of Neanderthals. For soon there would be a human genome with which to compare the Neanderthal results. Sequencing the Neanderthal genome was a superhuman effort and when complete, told us an amazing thing. Despite the vast gap in time since we humans split from the Neanderthals (i.e., about five hundred thousand years ago), our ancestors did mate with them, and the offspring were fertile. We are aware of hybrid animals, such as the mule, which is the product of mating between a horse and an ass. Mules, unfortunately are a dead end. Mules cannot reproduce. So, Humans and Neanderthals were genetically closer because those hybrids could reproduce. Eurasians alive today have about 2.5 percent Neanderthal DNA in their genomes. In fact, my son Kurt had his DNA tested, and he matched that value.

Paabo's team was not done surprising the world. They followed this amazing feat with the discovery of an unknown species of human, which they named the Denisovan. We have no idea what the skull or skeleton of this species looks like because all we have ever found was a finger joint and a tooth. Those fossils were rich in DNA, and that DNA was well preserved in the Siberian cold. We do know how this species is related to us and other humans. The Denisovan had common ancestor with Neanderthals 640,000 years ago and with humans 804,000 years ago. We also know that humans mated with the Denisovans and had viable offspring. This is evident in inhabitants of Papua-New

Guinea and the nearby Pacific Islands, where their genomes indicate a 5 percent contribution from the Denisovans.

Bioinformatics

A new science has sprung up that combines DNA science, computer coding, and biology to investigate genetic relationships using thousands of genomes available online. Thousands of human genomes, from around the world, already exist online, and more are coming. The potential of this new science is awesome.

Eight

Culture, Intelligence, and Innovation

The Chapter in a Nutshell

The multimillion-year age of the Australopith was a period of evolving survival skills in an environment that was foreign to them, on the ground where they had neither speed nor fighting ability against predators. Superior vigilance and coordination were selected for over the generations. The Homo lineage evolved from these roots about 2.5 million years ago.

The hunter-gatherer phase of human evolution was significant in terms of increasing intelligence. In fact, we see a phenomenal increase of the human brain during their time span. In contrast to the unchanging brain volumes of the Australopiths, the Homo-lineage brain grew to a size three times bigger. Larger skulls in turn, affected childbirth, the span of child dependency, and the social structure because the majority of brain development occurred after birth.

The *Homo erectus* period was a time of significant evolutionary change, where most of the features of modern man were acquired. *Homo heidelbergensis* continued the trend with his still bigger brain and mental capabilities. *Homo sapiens* existed in Africa from about three hundred thousand years ago and became the most successful species of Homo. Their secret was an increasingly innovative spirit, and it was most evident in the "out-of-Africa" migration some fifty thousand years ago.

Charles Darwin, Richard Dawkins, and others have credited sexual selection as being the major evolutionary force shaping human characteristics. Shapely breasts, hidden ovulation, and other traits are examples. Human mating has evolved to be a prolonged and pleasurable pastime unlike anything found in other species. Human females have learned to select male partners who are more supportive and less aggressive as child raising became more burdensome.

Speech is unique to humans; even our ape cousins are incapable of it. Some scientists believe that stone toolmaking is linked with the development of speech. Speech might also have arisen to enhance survival because it warned of predators, helped coordinate hunts, and helped pass critical skills like knapping forward to successive generations. Pair bonds between male and female hominins most likely formed during the Homo erectus period with the result that population size grew, and sexual selection enhanced traits to attract and keep optimal partners. Speech may have been a key factor in selecting the best mate. Genetics tells us something about speech development. The FOXP2 gene exists in most animals, but humans have a version of the gene sufficiently different to account for the fact that humans can verbally communicate and other animals, even apes, cannot.

Reviewing Our Evolutionary Past

From Tree-Dwelling Apes to Bipedal Apes

Our ape ancestors descended from the trees some five to seven million years ago because the tropical forests were disappearing due to prolonged climate cooling and drier conditions. These changes also continually reduced the supply of their traditional diet of fruits and nuts. Traversing treeless zones in the search for food made their lives riskier. As long as they stayed aloft in the trees, their superior speed and agility of swinging through the branches made them virtually immune from predator capture. However, fewer and fewer of them had that option because their forests were disappearing and were transforming into open woodlands. As the centuries went by, grasslands were separating the groups of trees more and more. They had to spend most of the daylight hours on the ground foraging and in the process of doing that became proficient bipedal walkers. This is one of the mysteries of our evolutionary history because no other mammal has committed to bipedal walking as did our ancestors. The baboons were once tree-dwelling monkeys, who also adapted to living at ground level but unlike us, they retained their quadrupedal mode of locomotion. During this bipedal ape phase, our Australopithecine ancestors retained their long ape arms and curved fingers, which helped them quickly scamper up a tree in the case of danger. Like their tree-dwelling ape cousins, they still made nests in the trees and slept there at night to be safe from the nighttime predators. However, they sacrificed some climbing ability as they evolved into better walkers. For example, their feet acquired better walking ability in exchange for the former ability to grasp branches. The big toe (i.e., hallux), once like an opposed thumb, now lined up with the other four toes. If we examine our own human feet and compared them to our hands, we can imagine the transformation. Numerous other skeletal changes occurred in the transformation of a quadrupedal ape into an erect-standing, bipedal Australopith.

The Emergence of the Homo Lineage

The Homo lineage first appeared about 2.5 million years ago. So, how does an anthropologist distinguish between a Homo fossil and an Australopith fossil? Well, you know what *Homo sapiens* look like. So simply put, you ask, "Does the hominin fossil in question seem more like us or like an ape?" We modern humans have distinctive features: we have large skulls, nonflaring rib cages, short arm bones, long leg bones, and so on. In contrast, the Australopiths have ape-size skulls, flaring ribcages to accommodate a big gut, short leg bones, and long arms. Modern hominins are a lot like us, so they are obviously assigned to Homo. But when we examine hominin fossils from around 2.5 million years ago, it is difficult to assign a genus because we are usually looking at a mixture of advanced and primitive traits. Moreover, the Homo lineage didn't just appear out of the blue one day. The lineage must have descended from one of the existing Australopith species, perhaps even one we have not discovered yet. Let's consider how that descend might have occurred.

The brain volume of the various Australopithecine species stayed about the same over their multimillion-year existence. However, brain volume tells us nothing about the composition of the brain matter. Those man-ape brains had been evolving to support a different kind of intelligence, where their senses and awareness were more acute, and where their interaction with others was more cooperative. The Australopiths developed superior survival skills for ground-level living than their tree-dwelling cousins ever had. They had keener vigilance. They saw more, heard more, and gleaned more information from their observations. These short-legged hominins were not fast runners, and so they needed as much advance warning time as possible to get to the nearest scalable tree. The cooperative teamwork of the troop made the critical difference. A dozen pairs of eyes can see more than one pair. A dozen pairs of ears can hear more than one pair. And a dozen brains processing these sensory inputs can detect lethal danger faster than one brain is likely to do. Teamwork, cooperation, and communication became more important to the survival of the Australopiths than it ever was for their tree-dwelling ape ancestors.

The cooling trend was millions of years in duration, and conditions worsened for our ancestors. As savannahs replaced the woodlands, the Australopiths faced a new set of problems. Climbable trees were increasingly more difficult to find, and the traditional diet of fruits and nuts was very hard to find. Survival depended on developing new food sources and better defensive abilities. One group, the newly evolving *Homo* genus, solved the nutritional problem by adding meat to their diet. Some anthropologists think that scavenging meat was the major method for acquiring it and that hunting came much later. They also acknowledge that scavenging had its risks. Lions, hyenas, and leopards would return to their kills and would make a meal of the hominin scavengers if they could. Stone-cutting implements made the critical difference. Our ancestors learned that stones, which fractured leaving a sharp-cutting edge, could help them get meat from the carcass quickly. They learned to hack off a large chunk of meat and get to a safe place with it before the predators returned.

No doubt, such scenarios occurred numerous times. The problem is that recent archeological finds indicate it wasn't scavenging that we see in the ancient sites but highly effective game hunting and butchering. Moreover, these slaughter sites are being discovered for earlier and earlier dates. The inference from one site is that hunting actually goes back to 3.4 million years ago. It may well be true that hunting was conducted by Australopiths long before the Homo lineage began.

As for survival on the savannah with its lions, leopards, and hyenas, huge physical changes occurred in *Homo erectus*, who mastered this environment. They developed bigger brains, longer legs, and an endurance runner's body, due in part to a superior cooling system for their bodies. These ancestors of ours had an ecological impact on their environment. They eliminated many of their competitors and rose to top predator themselves. They could run down game, not with greater speed, but by outlasting them. Their bodies evolved due to the meat-containing diet. They evolved smaller guts and a vertical ribcage to go with it. They also surrendered the long arms of the tree climber in exchange for shorter ones. They no longer needed the tree-climber's long arms because they learned how to live safely on the ground.

They tamed fire, and that may have been very important to discourage predators at night. They made sophisticated stone tools and weapons (i.e., the Archulean culture) and probably had speech and language. Speech could be vitally important because planning and coordination are possible, and valuable lessons can be passed down through the generations. Spears could be an effective weapon against marauders on the savannah, and thorny branches an effective barrier fence against them.

And yet, evidence mounts that tells us that we were becoming a top predator long before these skeletal changes occurred in *Homo erectus*. The fossil and artifact evidence for early Homo is scant. Yet, the pieces of the puzzle dribble in little by little. Somehow our very ancient ancestors were mastering their environment. I think that intelligence and teamwork must have been a big part of their success.

The Hunter-Gatherers

What is a hunter-gatherer? It is a lifestyle that combines hunting with harvesting whatever edible plants, you find along the way. Moreover, hunter-gatherers are nomadic to varying degrees. As nomads, they sometimes ventured into unfamiliar lands with unforeseen dangers. Survival under such a lifestyle would depend upon being resourceful and adaptive to unexpected events that spring up. Thus, high intelligence and cooperative behavior must have been traits selected for over the generations.

Instinctive behavior should not be ignored though. Doing the right thing, the instant danger appears can be vital to survival too. The hunter-gatherer mind-set has been with us for at least two million years and probably longer. It hasn't disappeared even today. There are hunter-gatherers still existing in areas of the world, although the ingress of modern civilization is deep as well. Now, much of our instinctive behavior was fashioned during this long hunter-gatherer period. For example, our fight or flight reactions still get triggered as hormones release adrenaline into our bloodstreams, increase our heart rates, loosen our bowels, and prepare us for an encounter with an enemy. Sometimes to our dismay, we experience them still. Yet, these days those reactions are inappropriate. Our human adversaries may get our dander up, but fleeing the scene or whacking him with a bookend is not acceptable behavior. Even as social groups today, our aggressive instincts get the better of us. In an age of nuclear weapons, biological weapons, and chemical weapons, survival depends on suppressing those baser instincts.

Civilization Emerges

The hunter-gatherer's reign was a very different time for humanity. A nomadic life meant that possessions were limited to essentials. If it was a burden to carry than it must be shed. Selfishness was unacceptable. If you had a surplus of meat, you must share it rather than let it rot. These attitudes changed when civilization arrived, and the transition may have begun as early as fifty thousand years ago with the great leap forward. The relics of this time tell us that innovation was being expressed at a rate unseen before. It never slowed down from that time forward. If anything, it accelerated. There seems to be a new awareness of one's humanity in during that time. It was expressed in decorated implements, cave art, figurines, jewelry, music, sexuality, and ceremonial burials.

The Agricultural revolution, which began around 10,000 BCE, was a logical next step in applying innovation to human problems. Game animals were overhunted, wild plants overharvested, as human populations increased. Farming and animal domestication were merely logical solutions to these problems. However, agriculture had a benefit no one anticipated. Populations grew at a rate never seen before. More people were staying put, and specialization of labor thrived. Of course, that meant even more innovation. The rest of the story is in the history books. Yet, one thing unmentioned in the history books is the continuing evolution of humankind. Selection processes now strongly favored higher intelligence, communicative skills, and ability to read the minds of other humans. The process continues today.

Sexual Selection, Human Mating, and Brain Development

Sexual Selection as an Evolutionary Factor

Human beings are unique among the animals in countless ways. Many of the distinctive traits that we have were not the result of natural selection, but of something else. It turns out that there is another powerful evolutionary force in nature, and it is called sexual selection. Many human traits arose because of it. The idea of sexual selection being an important evolutionary factor began with Charles Darwin, who wrote extensively about it. However, it didn't really catch on in the scientific community until recently. Many naturalists strongly associated evolutionary change with Darwin's concept of natural selection, where survivability is most important, but fewer have looked to see the evolutionary effect due to sexual selection. In fact, over a century went by before sexual selection was properly appreciated. The point here is that being good at surviving is only part of what makes a species successful. Growing their population is the other essential part. As species compete for the resources of the planet, they must continually produce high-quality offspring. To do this, the superior genes of the best males and females must be passed to the next generation. That is the role of sexual selection, and Darwin was fascinated in how different kinds of animals do this.

Sexual Selection in Birds

Take birds for example. Everyone knows that male birds are the more colorful and decorated gender. They also sing wooing songs to attract the females, and some even build elaborate nests in order to convince choosy females to select them for mating. Peacock plumage is often cited to show the power of sexual selection in altering species.

Figure 8.1 Male Peacock Displaying his Grandeur
Courtesy of Wikimedia Commons

Consider what the peacock looks like today. The courting peacock fans out his immense, elaborate tail, which is embroidered with beautiful colorful designs. The peahen has been selecting the male with the most ornate plumage for countless generations. That is why the peacock tail has become what it is, but it wasn't always that big and ornate. A long time ago, when a few birds had survived the K/T extinction event, the peafowl's ancestors were normal-looking birds with much smaller and plainer tails. For some arbitrary reason, peahens began selecting males for their extraordinary tails and shunning peacocks with drab, normal tails. This sexual selection tradition became embedded in their genes. Peahens were genetically programmed to select males with fancy tails, and the offspring developed fancier and fancier tails.

I'd like to present yet another example of spectacular plumage. When I was in my teens, I would look forward to when hunting season would open in November. As I recall these times, I picture ring-necked pheasants, felled Illinois cornfields, with their stalks broken and down, patchy snow on the ground, and the visible breath of my companion hunters. In those days long ago, I learned a few things about pheasants. For one, I learned that the male ring-necked pheasant is quite a beautiful bird, with its rich coloration and iridescent hues. However, I also noticed that the female is comparatively dull and artfully camouflaged. I further observed that the infant pheasants are so well camouflaged that you don't see them unless they move, and this is true of both sexes. So, nature keeps the male safely camouflaged until he reaches maturity and then adorns him in beautiful feathery splendor so he can compete with other males for his chance to father the next generation. By the way, I eventually quit hunting in disgust when it ceased to be sporting. For years, only a few hunters ever worked these cornfields. That paradise ended one year when hordes of city hunters invaded our rural farmlands like a conquering army. Sometimes humans behave in a contemptible manner.

I think that birds are fascinating, so let's do one more example of sexual selection in birds. This time the topic is the Galapagos Great Frigate bird. These large, dark seabirds have long-pointed wings and a deeply forked tail. Their short legs end in webbed feet, and their long beaks hook downward at the tips. They are bullies of the seacoast, stealing the fish caught by smaller seabirds. Here is how I encountered them. My family and I took an ocean cruise of the Galapagos Islands, which are six hundred miles west of Ecuador. Each day we were boated ashore and joined a walking/hiking group led by a tour guide, who would describe the points of interest of the particular island.

Figure 8.2 Galapagos Great Frigate Bird
By Charlesjsharp (Own work) [CC BY-SA 3.0 (http://creativecommons.org/licenses/by-sa/3.0)], via Wikimedia Commons.

One of the primitive islands that we visited, happened to have Great Frigate birds in the process of mating. This was really something to behold. The male birds had built nests, which would be inspected by the females as they flew into the wooded area. The nest builder also got appraised by the choosey females, and here is the spectacular thing about them: The male has a bright-red gular pouch under his beak, which he inflates when the females conduct his inspection. The inflated pouch adds significantly to the size of the male bird. This inflatable red pouch is the product of sexual selection at work over numerous generations. A peacock tail wouldn't work for a seabird, but an inflatable pouch seems perfect.

Those bright colors and displays of male birds may win fair maiden, but they also must wreak havoc with the male bird's safety. Although sought-after females may be watching, salivating predators are watching too. So, what is going on here? Why has nature converted the once-camouflaged male into a bright-colored target for predators? The answer is sexual selection. The fact is that females have more at stake than males do when the topic is reproduction. It is their bodies that produce a few precious eggs, whereas sperm production is cheap and plentiful by comparison. And because of that value difference, females get to choose who they will mate with. Moreover, they want the best male partner they can find. Consequently, each species makes those decisions based on a male trait, which was selected in the species' distant past and has intensified over countless generations. Hence, the peacock's ornate and grandiose tail, hence the cock pheasant's iridescent colors, the frigate bird's inflatable red pouch, the meadowlark's melodious song, and countless other examples.

Sexual Selection in Mammals

Mammals display sexual selection behavior too, but it often centers on male contests of strength and combativeness. The field of battle determines which male gets to pass his genes on to the next generation.

We see this vividly with grazing animals. A dominant male controls a herd of females. Most important, he has exclusive sexual access to his females, and he aggressively prevents other males from mating with them. Eventually, he will get older and weaker and then a younger, stronger male will take the female herd away from him. Nature equips those mature males for the contests. Male deer and elk grow antlers for these contests, whereas does and young deer and elk of either sex have none. The drive to mate comes with a cost. Sometimes competing males lock antlers irretrievably, and both of them die as a result. Male mountain sheep each take a running start and collide head to head with crashing horns in a contest for dominance. The victor of these brutal encounters gets to pass his genes forth on to the next generation. Just imagine the evolutionary process that step by step selected for ram bodies capable of surviving such violent encounters. Those colliding rams are the beneficiaries of superior fighting genes. Shock-absorbing tissues have become ever more efficient over the generations in allowing the winners to survive. Those genes enhancing shock-absorption were passed forward, whereas the genes of losers died with them.

Sexual Selection in Apes

Now, moving the discussion closer to our own species, let's examine the mating behavior of apes and see what we can learn from them. The gibbons and orangutans of Asia are not in our immediate ancestry but do contribute some genes to us. They are more tree bound than the African apes, probably because they have hungry crocodiles and tigers awaiting them on the ground. Gibbons tend to form male-female pairs for life, although those bonds do not extend to absolute sexual loyalty. Orangutans tend to live solitary lives, but males and females get together occasionally to procreate. Male orangutans come in two different versions. Let's call them small and large. Females do not want to procreate with the small ones but are sometimes raped by them.

The large males are the huge, masculine, auburn-colored ones, which we usually associate with the species, and the females do want them as mates.

Moving over to Africa, we have three main kinds of apes: gorillas, chimpanzees, and bonobos. The gorilla has a harem type of society, somewhat like the grazing animals. A dominant male rules over several females and their offspring. When a young male has matured to the point of being a competitor to the dominant male, he is driven out of the troop. Turning tree-ward, we see that chimps and bonobos are very similar-looking animals, but their cultures are quite different. Chimps and bonobos split off from a common ancestor about one million years ago, and the Congo River has divided and isolated them ever since. Both of them have an open sexual approach to reproduction. When a female demonstrates that she is in heat, males gather around her and take turns with the result that parentage is uncertain. The male apes do not fight each other for the female, but their sperms do. More sperms and faster-swimming sperms have the advantage in this kind of contest, and so these particular apes have the largest testicles of all the apes. Aside from these similarities, chimps and bonobos are quite different. Chimps have male-dominated societies, and aggression is part of everyday life. Bonobos, on the other hand, have female-dominated societies, and they are far less aggressive. Bonobos have substituted sex for aggression as a problem solver.

What Ape Sexual Cultures Tell Us

So, we can see that ape sexual cultures are all over the map: gibbons as monogamous partners; orangutans usually living in isolation but meeting up briefly to mate; gorillas in troops of females and offspring headed by a dominant male; chimps, male dominated, social with aggression, males taking turns with the females in heat; and bonobos, female dominated, social, and nonaggressive, where sex is their daily common social tie to each other. What does this diverse range of sexual cultures tell us about humans and their evolutionary history? Perhaps, this ape review is helpful to answering that question, or maybe it is unrelated. The vast majority of anthropologists believe we are descended from African great apes, so we can narrow the list down to the African apes; gorillas, chimps, and bonobos. Although we carry a good share of gorilla genes in our genome, we are nothing like a gorilla culturally. Gorillas can live safely on the ground because they are big and very powerful. We are not like that. Gorillas eat low-nourishment foods like grasses. We never went down that dead-end route. That leaves only chimps and bonobos to consider. Actually, we descended from the common ancestor of those two species, and we don't know anything factual about that ancestor's sexual lifestyle. We think it is likely that the common ancestor lived in trees, ate fruits and nuts predominately, and were highly social. Furthermore, chimps and bonobos look a lot alike after a million years of isolation, so we assume the common ancestor looked like them too.

Anthropologists look to sexual dimorphism to infer something about sexual culture. In other words, were the males a lot bigger than the females, or were the two sexes close to the same size? The former condition infers males disengaged from child rearing, whereas the latter condition infers more male involvement with child rearing. The Australopiths tended to display high sexual dimorphism indicating their males were disengaged. On the other hand, the Australopiths lacked the protruding eyeteeth (i.e., fangs), which apes possess. They probably were not using male-to-male combat to decide rights to mating. The Homo lineage reflects a significant reduction in sexual dimorphism, and we know for a fact that males had increasing involvement in child raising and family protection.

Sexual Selection in Humans

Human beings are animals regardless of what some people claim. We eat, drink, eliminate waste, breathe, procreate, see, hear, smell, and taste like other animals do. Yet, we feel that compared to the other animals, we are very special and perhaps not really just an animal. We don't have fur, we alone can communicate through speech or writing, and we alone can invent things to make our lives easier and safer. There are indeed many special things about us, but our distant ancestors were tree-dwelling apes, who had none of these special traits. Scientists that study this transition from ape to man tell us that sexual selection played an important part in affecting how we are the way we are. We will examine this process deeper next.

How Humans Evolved Sexually

The Influence of Sexual Selection

What does sexual selection have to do with us humans? Most scientists studying human evolution think it was a major factor in making us what we are. Humans are unique in the animal kingdom. We are different in our upright posture, and in our bipedal walking, running, skipping, and dancing. No other species has language and communicates verbally. We think about what we are doing right now, but with one eye to the past and another eye to the future. Our appearance is unique. We are mammals without fur, which is strange because we don't live in the ocean like furless whales, nor are we too large to cool ourselves like an African elephant. Our bare skin comes in colors ranging from pale pink to dark, dark brown. Brown eyes are dominant in the world, but blue, hazel, and green eyes are common too. Kinky hair, straight hair, and wavy hair decorate human heads in colors of black, brown, blond, red, and mixtures of these colors. Our women are unique with their full breasts, hidden ovulation, and ornateness. Our males are unique with their facial hair, muscularity, and large penises (compared to those of apes). Sexual selection had a hand in making us this way, but how far back did it start? Did it begin with the Australopiths? Surely, they developed cultures different from the tree-dwelling apes. The appearance of members of the opposite sex was different for Australopiths than it is for apes due to the changes in having an upright posture.

Females couldn't advertise their "in-heat" status with genital swelling because the genitals had moved to where they are far less obvious. The fleshy parts of Australopith remains are not preserved for us to examine, but sexual selection must have influenced their appearance. However, it is in the Homo lineage that sexual selection leaves its mark.

From *Homo erectus* to Modern Humans

We can learn more about our sexual evolution by examining the Homo lineage than we can from the bipedal apes because their time is more recent and because our own bodies display many of those changes. The *Homo erectus* fossils, tools, and other artifacts tell us that here was a species that was becoming more and more similar looking to us with time. Our ancient ancestor's skulls were differently shaped and their brain volume less than ours, but their skeletons were very similar to ours. The Homo lineage evolved bodies designed for running, which includes long legs, small gut, and arms much shorter than the Australopith arms. We also evolved an efficient cooling system consisting of densely populated sweat glands in our skin and a blood circulating system to cool our brains. The sweat-gland acquisition necessitated the drastic thinning of our body hair in order to work efficiently. These changes to become a runner happened during the reign of *Homo erectus.* Turkana Boy, who was discovered in Kenya, already had most of these attributes over 1.5 million years ago. However, there are sexual traits in today's men and women, which are not preserved in fossil bones.

Let's examine some of those uniquely human traits. One of them is prominent breasts. Today's women have prominent breasts from puberty on. All other mammals only have full breasts while they are lactating, so this is a curious trait. Scientists believe that permanent full breasts are a sexual selection trait, but we can't say for sure when that trait was acquired. Was it during the Homo lineage evolution period, or even earlier? Broad hips and narrow waists are other female traits that attract males. These traits of full breasts, narrow waist, and broad hips are thought to embody a subliminal message that this woman is fertile, capable of safely having babies, and capable of providing for them. Another unique human female trait is hidden ovulation. Female chimps and bonobos have the very opposite trait. They sport a swollen and highly colorful vulva when they are in heat, whereas human females display no indication of ovulating. We can't say when this trait first appeared in our evolutionary history, but it seems to be a key factor in the success of human pair bonding. If females are no longer advertising their availability for mating to the available males, one male can claim fatherhood of her offspring and take an interested role in raising his children. Committed male-female partnerships should provide a better environment for child raising than the one where women bear the full responsibility. The children might even be spaced fewer years apart. The end result would be higher population growth, and that is a vital index for the success of a species.

Human Pair Bonding

Once pair bonding became important in human culture, both males and females had incentives in selecting the best partner possible. Females of all species have a stronger incentive than males in selecting the best partner. They carry the embryo to term, bear the risks of childbirth, and then must feed and protect the infant for years. However, if the female can enlist a suitable male into making a similar strong commitment to child raising, her burden is eased markedly. This new cultural factor induced females to become more attractive sexually, so they could each find the best partner in terms of competence, reliability, and good genes. So it seems as far as humans are concerned, females no longer were the only ones doing the selecting. Males also became selective in choosing to whom they would make a long-term commitment. An inducement was the benefit of having an ongoing sexual relationship and willing partner.

Sexual selection arose from the successfulness of these human pair bonds. Males were attracted to females who provided sex to them exclusively on a daily basis. Hidden ovulation, perpetually full breasts, kissable lips, and other attractive female traits served to attract the best male possible into a committed relationship. She provided fidelity to that special male. He, in turn, felt confident that their offspring were his offspring, and he became more motivated in providing for them and protecting them. However, sexual selection was not only transforming human females, males developed notable traits too. Most remarkable of them is the human male penis. The typical ape penis is small and has the girth of a pencil. They also contain a bone and have a limited range of motion. The typical human penis is boneless, flexible, markedly longer, and thicker and is not as quick to ejaculate. How did it transform itself from the quite different ape version? It was transformed due to female selection. Females were selecting based on what felt best to them and brought them the most pleasure. Once sex became an entertainment device to cement pair bonds, human females wanted the best they could get from this entertainment device too. Pencil-sized penises and quickie encounters were not it. That is the way sexual selection works. It was peahens that fashioned elaborate peacock feathers, hen pheasants that fashioned the colorful male pheasant's plumage, and human females that fashioned the sexual equipment of and lovemaking ability of human males.

Fatherhood

Perhaps this is a digression, but I was just struck by the very human trait of fatherhood. We have all been around long enough to observe different kinds of fathers. Some are even more loving and more involved with their children than are the mothers. On the other extreme, some fathers care little about their offspring and may even abandon them. Some fathers love their children but express it by working long hours to assure they have food, shelter, security, and an edge in a competitive world. Unfortunately, longer working hours usually means less time with their children. Perhaps apes are not devoid of fatherhood instincts, but it seems quite different with them. The male gorilla protects his females and offspring with his life if need be but expels young males from the group when they are mature enough to be seen as competition. Of course, chimp males never know if any of the infants are theirs, so they lack a strong sense of parenthood. At any rate, male chimps are not known for being very involved with raising infants. From what we know from human-recorded history, fathers from the earliest times have deeply loved their children. Different cultures generate different attitudes to be sure. In cultures, where they are struggling at the edge of daily survival, the priorities differ markedly from wealthier, more secure societies. Moreover, cultures are continually changing. My dad employed the old European attitude of a near-dictatorial control over his family. It was clear that I had better not disgrace him or else. On the other hand, he was often a friend and a mentor to me. I was a bit freer with my children I hope. They have been and still are the joy of my life. The point of all this is that human bonds are immensely powerful forces, which go back far into our existence. The strong emotional aspect of human fatherhood may very well be another product of sexual selection.

Human Intelligence

Inheriting a Primate Brain

Humans can easily claim to be the most intelligent animals on the planet. We can do things that no other animal can, like safely dwell in comfortable homes, travel across the country in a few hours, and get news of events around the world as they happen. Intelligence is certainly a big factor in accomplishing what we have. We have the highest brain volume per unit body mass of any creature that ever lived. And yet, like all other animals we are connected to LUCA in an unbroken line of descent. We have to wonder how we came to be so special. Did we inherit our unique abilities? So what is it about our evolutionary history that makes so smart? It seems like luck had a role in it. Our first lucky break was being a mammal rather than a cold-blooded reptile. The reptilian brain was a dead end in terms of intelligence. We still have that primitive reptilian brain, and it runs our basic life processes, but we mammals have a neocortex too. The neocortex added higher order functions to our reptilian brains. These include sensory perception, cognition, generation of motor commands, and spatial reasoning. Our second lucky break was being born a primate. Primates move through a three-dimensional world of tree branches where falling could have fatal consequences. Thus, vision and grasping ability became critically important, but good judgment became critically important as well. Primates have the highest density of neurons of any mammal due to these special needs. Third, DNA comparisons tell us that we descended from African apes that lived five to seven million years ago, and that is fortunate because apes are significantly more intelligent than their monkey ancestors. We can next trace our mental growth through those particular apes that came down from the trees and mastered bipedal walking. They are the Australopiths.

The Australopiths descended from tree-dwelling apes and for nearly three million years thrived as bipedal walkers, spending much of their time on the African ground. One species of them led to the Homo lineage that led to us. What can we say about the mental growth of Australopiths? We know that their brain volume was on a par with existing apes, which casts doubt on whether they were any smarter than tree-dwelling apes. I think they were smarter though. For one thing, their lives were more at risk from predators than when they lived in trees. Survival meant spotting danger soon enough that they could get to safety in the nearest climbable tree. Many eyes work better than fewer eyes, so surveillance became a team effort. Brains adapted to this coordinated intelligence and interdependency. Second, Dean Falk's investigation of the brain endocast of *A. africanus* showed evidence of evolutionary improvements had occurred over ape brains. Third, recent fossil finds make us rethink the importance of brain size in intelligence. The Hobbit of Flores Island is of the Homo lineage, made stone tools, and hunted and yet only had an ape-sized brain. *Homo naledi* of South Africa lived contemporary with big-brained Homo lineage species and yet had a much smaller brain. There is strong evidence that they buried their dead within a dark, torturous cavern. Prior to this contrary evidence, we didn't think that it was possible for a small-brained hominin to make stone tools, hunt, or bury their dead. Finally, there exists evidence of hunting and butchering of grazing animals over three million years ago supposedly before Homo species evolved. This makes us reassess what the Australopiths were capable of doing. They have been a lot smarter than we credit them.

Brain volume stayed pretty much the same during three million years of Australopith life but underwent a phenomenal growth during the time of the Homo lineage, beginning some 2.5 million years ago. The brain volume of the fossils from that earliest time fell in the range of 600–700 cc, which exceeds the typical brain size for the Australopiths. That value grew to 900 cc 1.6 million years ago to 1,100–1,200 cc six hundred thousand years ago, and to our current average of 1350 cc some three hundred thousand years ago. Our human brains are three times the size of an ape's brain. This unusual growth rate makes us more than a little curious as to what was driving the change. Numerous explanations have been proposed for this including the mental demands of stone toolmaking, sexual selection, the increasing complexity of thought and speech, and social-group dynamics. Possibly all these factors made more intelligent brains evolve. Here is another curiosity to ponder: our species has had our large brains for around about three hundred thousand years, but they first showed a spurt of creativity fifty thousand years ago in what is called the "Great Leap Forward." Why did it take so long? The great leap forward was a period of constant innovation and discovery. We see continuous improvements to stone toolmaking techniques, sewn clothing, invention of the spear thrower, and later, the bow and arrow, fishing hooks, barbed spears for fish and birds, warm dwellings, and long-distance trading. For the first time, we see art and music proliferate. Religions took root, and ceremonial burials showed life after death was in their minds. These innovative trends accelerated with the agricultural revolution and the rise of civilizations. Our human brains were evolving rapidly to accommodate such rapid change from the simpler hunter-gatherer life.

Neural Plasticity

There is a lot we don't know about the human brain, but modern science finally has the tools to investigate it properly. Using tools like magnetic resonance imaging, we can see the parts of the brain that light up when we are presented different tasks. One of the things recently learned is that brains of individuals vary more than their physical bodies do. The idea that brains are all made to the same detailed plan is wrong. There seems to be a lot of plasticity in how the cortex of brains develops. Moreover, we retain that plasticity through one's lifetime. For example, individuals born blind shift the visual part of the brain toward better utilization of their other senses, such as hearing. Many blind people learn to use echolocation to determine where they are and to navigate the space.

Skoyles and Sagan tell us in their book *Up from Dragons* that doctors see this neural adaptation in the case of amputations. For example, a person with a missing finger may feel the finger nerves respond when they are touched on the face. Many other examples of adaptations in the brain exist. Now that we are aware of neural plasticity, we are already employing it in the field of medicine. Doctors are beginning to use this discovery to treat injuries and birth defects. Wheelchair patients in Germany became independent walkers using neural plasticity to train the spinal areas in walking exercises with hoists and treadmills.

Neural plasticity explains many things about human evolution too. The Australopiths reorganized their brains to enhance their senses for seeing and hearing but particularly improved the processing of that sensory input to keep them safe. I suspect that rudimentary verbal communication was learned and used to communicate danger. The capacity for trust and cooperation must have increased too. These traits were also important when the shift from vegetarian to meat-eating nomads took place. This is the time of *Homo erectus*, who became a better and better hunter-gatherer in the African savannah. However, they were not strictly an African people but had occupied Asia and Europe as well. This took extraordinary problem-solving skills. Adapting to variable climates, new terrain, new predators, and a host of unforeseen problems required good decisions and fast adaptation. It would seem that verbal communication would be of great value here. It was our neural plasticity that made that verbal skill set possible. Now it is hardwired in our offspring. They learn to talk and comprehend within a year or two.

Communication

Human Communication Today

How do humans communicate with each other? Today's humans engage in speech and gestures, written communication, and even reading subtle facial signals at a very complex level. We even attempt to interpret what isn't said. It is called reading between the lines. What is he or she really trying to tell me? Communication is not unique to humans, but we center our lives around it. Moreover, it is hardwired into our genomes so that when babies babble and coo, they are already practicing to speak. Speaking must have been critically important to our ancestors because our young children learn the basic skills so easily and so rapidly. The average four-year-old already knows five thousand words; by age eight, he/she will know twice that number; and as an adult, he/she will know at least twenty thousand words and perhaps as many as thirty-five thousand words. However, we can do a lot more than just understand words, we can sense whether the words are expressed seriously or in jest, whether the words comprise an honest message or a deceitful one, and whether the speaker is to be trusted or feared. We have these skills because we spend most of our youth learning them in schools and practicing them on our peers. And society wants us to get even better at it than we are. Within my own lifetime, I have seen the minimum level of education advance from a grade-school graduation for my grandfather to a high-school diploma for my teen years, to a college degree or even an advanced degree today. That trend is likely to continue. Future children might be pursuing doctorates at an unprecedented rate.

The Power of Speech

We are the only animal that can communicate verbally through language. Apes are smarter than monkeys, but even apes are incapable of speaking words due to physical limitations as well as genetic ones. Chimps and gorillas have been trained to communicate simplistically with human coaches using sign language, but a two-year-old human child is already far more capable of using words than the smartest ape will ever be. Speech and comprehension of it are hardwired into humans as we observe in human children learning the skill with or without our encouragement. Speech lets us communicate plans, complex thoughts, bond with other humans through gossip, educate or direct human activity, and finally to pass information from one generation to the next. Moreover, as our vocabulary builds through life, we are more and more capable of complex thought.

When Did Human Verbal Communication Get Started?

Speculation

What would we observe of humankind if we could run the clock backward and see simpler and simpler times? In fact, let's go way back. When did humans first learn to speak and comprehend even one- or two-word sentences? The experts disagree with each other on when we first used speech. It comes down to speculation in the end. It would be immensely useful to know when humans first used verbal communication routinely. The rest of human history would suddenly make a lot more sense. As it is, there are a range of expert opinions on that question ranging from as far back as during Australopithecine times to as recent as a few thousand years ago. Let's see if we can do better than mere speculation.

What Can We Learn from Stone Toolmaking?

One school of thought is that speech may have existed about the time when stone tools were being first used to cut meat from animal bones. The tradition of stone toolmaking passed from generation to generation so there must have been some way for the technology to have been communicated across generations. Sign language is one possible way, but verbal language would have been more precise. So how far back was that? There are a number of sources that tell us that the first use of stone tools by prehistoric man goes back 2.5 million years. However, new evidence pushes that date back even further, to about 3 million years. The stone tools are a surviving relic of those times, and anthropologists have assigned a name to these first stone tools, the so-called Oldowan culture. You would recognize them as very primitive tools. In fact, they can be as simple as a rock fragment with a sharp edge, but here is what makes them important. These stone tools have been found at sites together with animal bones with intentional cut marks on them. These cut marks are what you would leave if you were removing meat from the bone. Only hominins could have done this butchering, and all these complex activities can be best explained if there had been verbal communication between the participants.

As mentioned previously I wonder if some kind of primitive verbal communication didn't originate even earlier in time with the Australopiths. Survival on the ground was a tenuous situation in light of fast-moving predators easily able to run them down. They did survive, and so we wonder how they did it. Verbal warnings conveying the type of threat and location of the threat would have made a huge difference to their survival rates. Natural selection would select for those groups with the better surveillance and warning skills over those with the poorer survival skills.

Whether you accept the argument that early stone toolmakers could verbally communicate or not, the thesis is more convincing when hominins achieved sophisticated stone-making skills. By 1.6 million years ago, we can see planning and workmanship in these beautifully fashioned stone tools.

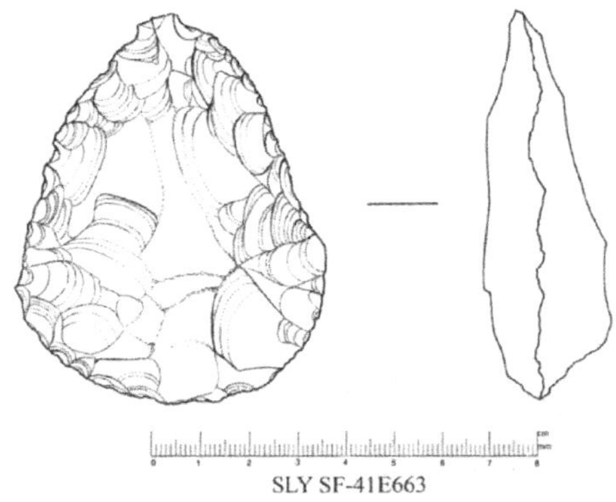

Figure 8.3 Paleolithic Hand Ax
Attribute: The Portable Antiquities Scheme/ The Trustees of the British Museum [CC BY-SA 2.0 (http://creativecommons.org/licenses/by-sa/2.0)], via Wikimedia Commons.

These tools are called the "Archulean culture." This toolmaking tradition is well documented from numerous relics found over many thousands of years. In particular, Archulean hand axes have been found in the hundreds from these ancient times. They are symmetrical tear-shaped tools with bifacial edges. Obviously, planning and careful shaping of these hand axes were required to make them. Even those among us, having good craftsmanship skills would probably not be able to make an acceptable one if we had a week to spend experimenting. In other words, our primitive ancient ancestors appear to have been better at this task than modern humans are today. Moreover, it is a fact that this skill-demanding procedure for producing hand axes was passed down through the generations because they appear at sites separated by hundreds of thousands of years. So, we might well consider that some kind of verbal communication was very likely in use back this far in time.

The Neanderthals of more recent times are identified with the Mousterian toolmaking culture, and it is considered to be more advanced than the Archulean culture. *Homo sapiens* used similar stone-making techniques but unlike the Neanderthals, surged ahead with a series of innovative changes. The great leap forward of the Cro-Magnon people in Europe was one rich in symbolism. Spear throwers and other items were beautifully carved to decorate them. Symbolism appeared in polychromatic cave wall art, sculptures, jewelry, musical instruments, and ceremonial burials. It is certainly true that complex verbal communication was in use by our ancestors during this time (i.e., about thirty thousand years ago).

Figure 8.4 Painting of Prehistoric Bison in Altamira Cave
By Rameessos (Own work) [Public domain], via Wikimedia Commons

Pair Bonding

A good case can be made that verbal communication, however it began, developed rapidly due to human culture. Apes bond by picking ticks out of another's fur, but humans bond through conversation. In mate selection, conversation is of utmost importance. Now, the normal relationship between the sexes in our day and age is to form a marriage or equivalent type of committed relationship between a man and a woman. This is especially the case if the couple intends to raise children. Our history books tell us this type of relationship has been the norm as far back as records go. So when did this practice begin? How far back is the advent of human pair bonding? Our ape cousins do not practice this type of male-female relationship, but at some point in our evolutionary history, we humans adopted it. It is an important question because committed relationships have made us more successful than the apes. All the living ape species are endangered in large part to their poor reproductive record. In sharp contrast, human population growth rates are explosive. Pair bonding has given us the advantage of doubling the number of committed parents, which allows for more children per couple and better care during their vulnerable years. Now let's add this information to the question of brain growth and verbal communication.

Three Traits Combine

The three human-defining traits of brain expansion, verbal communication, and pair bonding appear to be all closely related and here is how: Throughout the long evolutionary history of the Homo lineage, the brain has been increasing in size. Eventually, the fetal skull became so big that it could not get any bigger, and the mother survives childbirth due to the limited size of the female pelvic opening. If the opening were any wider, the angle of the female femur would be too extreme to allow her to walk. So once that maximum head size was reached, all further brain growth had to be outside the womb. That extended brain-growth period is also a period where the baby remains highly dependent. Today's human babies are helpless for an abnormally long time compared to other animals. They require constant watching and protection. By contrast, chimp babies mature much faster and can get around far quicker. Consequently, human mothers have an extraordinary task during the first two or three years after childbirth. To feed themselves and care for a highly dependent baby in a hunter-gatherer society would have been difficult to say the least. These arguments lead us to believe that pair bonding began during the days of *Homo erectus*, when the demand for greater childcare was increasing. If she had a male provider and protector, the burden is lightened immensely. Both male and female have to feel right about their partner if such arrangements are going to last.

From the male's viewpoint, sex is what probably drew him to this female, and if it is an ongoing thing, all the better. If she prepares meals and shares with him, that is better than fending for himself. In time, possessiveness and affection might develop. Ongoing sexual relations result in pregnancy and children. He is unlikely to abandon his own children, his woman needs assistance, and a sense of responsibility develops.

From the female's viewpoint, selecting the best male mate was a critically important decision. For one thing, compared with today's humans, our ancient ancestors tended to be more violent and more impulsive. The pacification of humans has been a long slow process, and even after all that mellowing, we are still quite dangerous. She wanted good genes from her mate, but she vitally needed assistance in feeding and caring for her children. So, successful, capable males would be high on her list. However, she needed to select a mate who would not kill or injure her or her children in a fit of rage. This is the point where speech becomes so important. Talking to prospective mates would be an excellent way to help her make that important assessment. If such a process of mental assessment was part of human sexual selection, then it helps explain the phenomenal brain growth during *Homo erectus* times. Individuals with greater intelligence would be a lot more likely to pass their genes forward into future generations.

Genetic Evidence of Human Speech

Perhaps, genetics can tell us something about when human speech began. What if there were a gene for verbal communication? Actually, there is such a gene, and it is called the forkhead fox P2 gene or simply the FOXP2 gene. Often, physicians learn about genes and their functions while investigating unfortunate humans who have functional problems and so it is with speech. Humans with a defective version of this gene have great difficulty forming words and speaking. Principally, they have trouble coordinating their mouth and facial muscles to properly articulate words. The gene has been identified in mammals as different from us as a mouse, but humans have an evolved version of it. In fact, our version is even different from the version found in our closest relative, the chimp. That important fact gives us confidence that the FOXP2 gene is the key element in controlling speech because humans have the evolved gene, which makes us capable of speech, whereas chimps with the older version of the gene are incapable of speech.

A precise date for the origin of the modern human version of the FOXP2 gene has not yet been determined, but it is thought to have originated within the last two hundred thousand years. Others have commented on the coincidence of the estimated gene age and the estimated age of our species being the same? However, I see little significance, and that coincidence seems to have been temporal because the age of our species has just been pushed back to three hundred thousand years due to discovery of a human fossil in Morocco. The appearance of the human version of the FOXP2 gene may represent a leap forward in verbal communication for our more recent ancestors, but I believe that human speech goes back far earlier than two hundred thousand years. Primitive speech may be linked with our rise to top predator much earlier in time, perhaps as long as three million years ago. After all, speech is affected by a lot more than a single gene.

On the other hand, the human version of the FOXP2 gene may be very significant in putting humans on a course for using innovation to solve problems of survival. If it evolved two hundred thousand years ago, it would still take tens of thousands of years to become dominant in the human population. The increased innovative spirit and symbolistic expression associated with the great leap forward fifty thousand years ago could easily be due to increasing frequency of the modern

FOXP2 gene in the population and the survival advantages that improved verbal communication would bring.

Before leaving the topic of the FOXP2 gene, I would like to mention an interesting experiment conducted by Wolfgang Enerd's team at the Max Planck Institute. Humanized mice were produced by implanting the human version of the gene in mice embryos. The scientists not only detected notable changes in the vocalizations of the altered mice, but their articulations had changed as well as their understanding of words. They concluded that significant and widespread changes had occurred in the parts of the brain concerned with verbal communication.

Modern Humans

We have seen how sexual selection is a powerful evolutionary force in nature and has played a major role in human evolution. We survived this long due to our intelligence and adaptability. Verbal communication gave us an advantage over all other animals. Our ancestry consists of over two million years of a hunter-gatherer culture, where innovative change was glacially slow. However, the great leap forward beginning fifty thousand years ago was a time of high innovation. We start to recognize our human traits in these ancestors. These are an appreciation of art, music, and expansive thought. In the next chapter, we will see how agriculture transformed the human world, the power of religion, and the rise of civilization.

Nine

Agriculture and Civilization

The Chapter in a Nutshell

These first experiments with a new way of living occurred in a fertile area in the Middle East. The Fertile Crescent was a rich area for farming due to the alluvial soil and the available water of the Tigris, Euphrates, and Nile Rivers. It was here that farming began, and it was from here that agriculture eventually spread westward to the Atlantic Ocean and eastward to the Pacific Ocean. Agriculture and animal domestication brought a surer food supply but brought infectious diseases from animal proximity and malnutrition from an overreliance on grains. Eurasians gradually evolved immunity to diseases and learned to eat a balanced diet.

Farming also developed in other parts of the world. Asia inherited the farming principles learned in the Fertile Crescent but went on to domesticate millet and rice. Asian agricultural techniques eventually were transported through the Pacific Islands. In the Americas, pastoralism was limited to the use of llamas and similar animals in the Andes. However, this usage never spread to other parts of the America. As a result, Native American farmers never had the disease problems that Europeans experienced, but they never developed immunity to these diseases either. So when Europeans invaded the

Americas in the sixteenth and seventeenth centuries, Native American populations were decimated by smallpox, measles, mumps, chickenpox, and other European diseases.

Mesopotamia lies between the Tigris and Euphrates Rivers in today's Syria and Iraq. This area would become the cradle of civilization as the Agricultural revolution took hold there beginning in 10,000 BCE. A stable food supply allowed city-states to arise, but it was strong religious convictions that united the people of these city-states and allowed them to work on projects for their common benefit. What began as small villages grew to the size of city-states with tens of thousands of people. Religious temples called ziggurats were erected at city center. Some were multistoried, elaborate structures requiring thousands of man-hours. Statues of their gods adorned the ziggurats and had an influence.

The Agricultural revolution fostered progress on many fronts. Stone toolmaking entered a new age, called the Neolithic Culture, where honed blades and pottery made their debut. However, it gave way to the copper age, where this malleable metal found multiple uses. Bronze is a copper alloy with broader usefulness. It had its own age too. Spear points and swords could be cast in molds. The discovery of iron processing ended the Bronze Age especially in terms of a metal for battles. Writing was invented five thousand years ago with huge implications for human advancement. The wheel was another powerful invention with benefits for human ever after.

Human endeavor spurred on by religious beliefs and powerful rulers ushered in an era of monument building across the world. The great pyramids of Egypt, the Sphinx, the Great Wall of China, the Parthenon in Athens are just a few examples of this phenomenon. Masterful planning, execution, and vast expenditure of human resources are evident in these great accomplishments. Considering that these people descended from nomadic hunter-gatherers only a few thousand years previously, the transition to architects and builders of such bold undertakings is breathtaking.

The rest of the chapter briefly closes the gap between these early civilizations and our world of today. Religion, technology, cultural progress, and violent conflicts seem to be the major forces shaping human destiny.

Prelude to the Agricultural Revolution

The modern humans who migrated from Africa into the Middle East and beyond were a more innovative people than had been seen in the world before. They were continuously finding new and better ways of doing things. Innovation paid off for them too. Their populations increased, and they had time for leisure and artistic pursuits. Especially in western Europe, we find polychromatic cave paintings that speak to us across a time gap of thirty thousand years. By 10,000 BCE, the multimillion-year-long cooling trend ended, and a warming trend began. This opened the door to the agricultural age. Some of our ancestors were using their ingenuity to take advantage of this opportunity. These innovators were planting wild grain and learning how to become the first farmers. Others had taken to pastoralism, the herding of sheep, cattle, pigs, or goats. Up until then, all humans had existed as hunter-gatherers. However, they realized that their game populations were diminishing and that wild plants alone were inadequate to feed their swelling populations.

The Innovative Spirit

Farmers

It was a radical change to go from the nomadic life of a hunter-gatherer to the stay put life of a farmer. After all, if you went to the trouble of cultivating the soil, seeding the ground, watering the seedlings, and weeding to remove competing plants, you were not going to let someone else reap the harvest when the plants were grown to maturity.

These farmers also had to think very differently than they did as hunter-gatherers. For example, they had to acquire the discipline to delay rewards. The seed for next season's crops must be preserved even when food got scarce. Another radical change in behavior was for them to think selfishly. Traditionally, hunter-gatherers generously shared their kills with the group. After all, the meat that was not eaten would spoil anyway. It was a better philosophy to share it, and perhaps the favor would be returned in the future. Farmers have to think differently. If they shared their harvested grain with strangers, there would nothing to eat in the months ahead. Generosity was no longer a good strategy. In terms of inventions, the need for new types of agricultural tools led to the Neolithic revolution. Stone toolmaking technology, which had previously been directed to hunting implements, was now directed to making farming tools such as scythes and hoes instead of arrowheads. One of the aspects of the Neolithic culture was the process of polishing stones rather than just chipping them. Grain harvesting and storage technologies had to be developed. The hard lessons of rodents eating your grain or humidity ruining it caused rodent-proof and water-tight chambers to be developed.

Pastoralists

Pastoralism is the raising of livestock. The animals were goats, sheep, cattle, pigs, and others. The farmers may have needed the most fertile land they could find, but the herders could feed their herds on the grasses of less fertile land as long as they could move on to greener pastures once the current pasture was exhausted. The farmers were bound to their farms, but the pastoralists were free to move their herds from place to place. In fact, a nomadic life was natural for these early herders. It was necessary to continually seek fresh pastures for their sheep, goats, or cattle. One thing that they had in common with the farmers though was the need to delay gratification. The temptation to feed themselves by slaughtering one of their herd, had to be tempered, or they would soon be without a herd. The wisest course would be to let the herds multiply to the largest size they could handle, then an occasional slaughter would less impact. Some herders found a way to grow the herds and feed their families without having to kill one of the herd. They were able to eat regularly by drinking the milk of their livestock. Milk was a rich source of food for daily survival, yet taking it did not diminish the herds. There was a biological obstacle to this strategy though. For most of the people in the world, nature shuts off their ability to digest dairy products about the age that they are weaned. The story might have ended here except some individuals, who by genetic accident, were able to digest milk into their adulthood. Such was the advantage of healthy bodies resulting from a milk diet that the number of people having this abnormal gene grew faster over the generations than those who remain lactose intolerant. Eventually the lactose-tolerant gene was dominant. The situation in the world today is that lactose tolerance is common in areas of the world where milk cattle are found, such as the Middle East and Northern Europe but remains uncommon everywhere else.

The Fertile Crescent

The Dawn of Agriculture and Its Effects?

The Fertile Crescent is also called the Cradle of Civilization. It was in this area that farming and pastoralism got their start about twelve thousand years ago, and it is where humanity began to change in a dynamic way. Remember that for over two million years, our ancestors had survived as hunter-gatherers. So, this new lifestyle put humanity on a radically new path involving a different diet, different daily life, and different relationships. The Ice Age was over with the result that climates were warmer and moister than they had been. Plants flourished and our ancestors noticed that certain edible plants grew seasonally in the fertile soil of the region. Over time, they learned how to select the best seeds and experiment with mixtures to produce the most ideal crops. Similarly, domestication of sheep, goats, pigs, and cattle was accomplished by selecting for desired traits in their wild ancestors.

The agricultural revolution was a mixed bag of benefits and liabilities, but overall, the benefits won out. No doubt the most important factor was that starvation was no longer the threat it had been in the past. Surplus food was produced through farming, and it could be stored to ensure there was something to eat in the lean periods. Another advantage of farming was that it was possible to settle down and stay in one place. Property was no longer limited to what you could carry, with the result that owning property took on a whole new meaning. Greater food production led to larger populations. People were able to specialize in a variety of pursuits and trading with other areas enriched the villages. Villages grew into cities, a variety of cultures emerged, and for the first time, the seeds of civilization were sprouting.

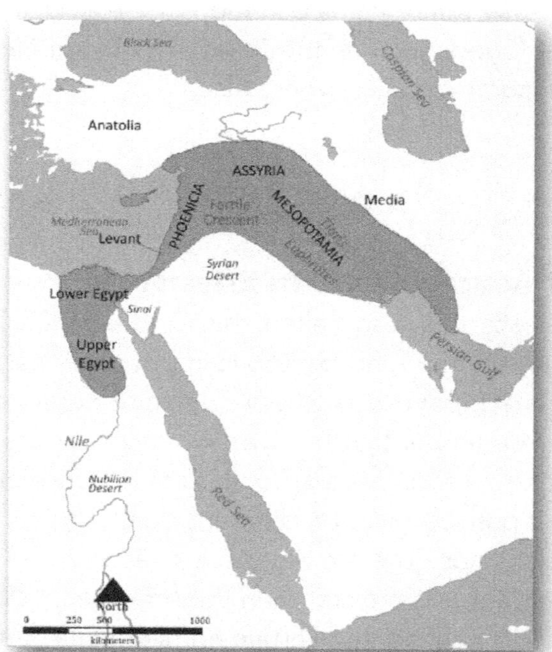

Figure 9.1 The Fertile Crescent
Courtesy of Nafsadh—Map of fertile cresent.png, GFDL, https://commons.wikimedia.org/w/index.php?curid=15272124.

What Is the Fertile Crescent?

The Fertile Crescent is as the name implies, was an area where crops once grew easily. It was here that farmers discovered that plants could be cultured toward human preferences through selection and cross-breeding. However, the Fertile Crescent is surrounded by vast zones of infertile land. Figure 9.1 shows the Fertile Crescent region that was so important to the birth of agriculture. The eastern end of the Mediterranean Sea is part of its western boundary. This immediate land area is called the Levant, and it includes Palestine and Lebanon. The fertile zone formed by the Nile delta of Egypt defines the southwestern end of the Fertile Crescent. From the Mediterranean boundary eastward to the Persian Gulf, the Fertile Crescent includes what is today Iraq, Iran, and Syria. Two great rivers flow through this region, and they have provided an alluvial plain, which is rich for agriculture. Those rivers are the Tigris and Euphrates. The northern edge of the crescent is formed by the foothills of the Zagros Mountains. The southern border of the Fertile Crescent is a hostile desert (i.e., the Syrian Desert).

How Did Agricultural Development Progress in the Fertile Crescent?

It is thought that farming began when the yield of edible wild plants proved inadequate to support the growing population. Perhaps, raising crops began after some people noticed that spilt grain often developed into sprouting plants where it had been dropped after a few weeks. By 9000 BCE, wild cereals, such as wheat and barley, had been successfully grown as crops. Farmers also experimented with crossing different plants to improve the grain size, yield, and other attributes. A domesticated wheat was introduced by 7700 BCE, and organized farming was underway in Egypt before 7000 BCE. Farmers also learned to solve farming problems such as long, dry seasons through innovation. For example, they were irrigating their fields by 5000 BCE.

Concurrent to the development of farming, the pastoralists utilized the less ideal land for herding of domesticated animals. We know that herders were managing wild sheep in the Zagros Mountains by 9000 BCE, and within two thousand years, those sheep had become domesticated. Moreover, the number of sheepherders was much greater. Goats had been domesticated by then too. By 6500 BCE, pigs were domesticated, and cattle were domesticated five hundred years after that. Horses were domesticated in Mesopotamia by 2000 BCE and were used for pulling carts and chariots soon after.

Agriculture Development in the World

General

As Professor Jared Diamond tells us in his great book *Guns, Germs, and Steel,*" the direction of agricultural spreading from its center of origin made a huge difference to its success. In particular, the east-west major axis of Eurasia gave it a significant advantage over Africa and the American continents, where their major axes lie in a north-south orientation. The east-west spreading of agriculture across Eurasia was both faster and more successful than the north-south spreading of agriculture in the other continents. The reason is that plants are adapted to thrive in the latitude where they are found and are maladapted for other latitudes. This simple concept is very important to understanding the history of the world during this transformational time. Crops can be successfully transplanted at the same latitude but are more likely to fail when transplanted along a longitudinal line.

Europe

The farming techniques, which were learned in the Fertile Crescent, traveled both eastward and westward. By 3500 BCE, farming was spreading across Europe. Dogs were domesticated in Europe by 10,000 BCE. By 1500 BCE, pastoralism was spreading across the Eurasian steppes.

Asia
Agriculture appeared in Southwestern Asia from 9000 to 7000 BCE in the form of wheat and barley. Sheep and goats were domesticated as well. The Yellow River Valley became an area of early farming for Northern China by 6000 BCE. Millet was one of the main crops. By 5000 BCE, rice, which may have originated in India, was widespread through Asia. Terraced farming was developed, and villages began to spring up along the Yellow River.

India
By 3000 BCE, the water buffalo was domesticated in India and became an important draft animal. By 2500 BCE, grain agriculture formed the basis of the Harappan civilization in the Indus River Valley, which is in present-day Pakistan and India.

America
Mesoamerica is a region extending from central Mexico southward through the central American countries of Belize, Guatemala, El Salvador, Nicaragua, and Costa Rica. It was here that farming got its start in the Americas beginning around 7000 BCE. Domesticated crops included beans, corn (maize), cassavas, squashes, potatoes, and peppers. Corn was cultivated in Mesoamerica by 2700 BCE. Meanwhile in the Andes of South America, we find potatoes grown by 3000 BCE. This crop will have a huge influence on the rest of the world after the conquering Spaniards brought it back to Europe. The alpacas and llamas were domesticated by 3000 BCE, but they never made it to central and North America.

Comments on the Meaning of the Word "Domestication"
We all assume that we understand what the word "domestication" means. In one sense it means that the thing is tame, which is the

opposite of wild. For example, the pigs on a farm are tame or at least domesticated, whereas wild boars are not. In general, we do not fear domesticated animals, whereas their wild counterparts may tend to do us real harm. Think of dogs, which are man's best friend, as opposed to their ancestral stock, wolves, which are not friendly in the least.

Now, the word "domestication" also infers that a change or conversion has occurred. Moreover, it is human intelligence directing that change to take place. Charles Darwin referred to this process as "artificial selection" to differentiate it from natural selection. Humans, not nature, are intentionally changing plants or animals to suit their own purposes. One of the applications of artificial selection was the custom breeding of pigeons, which was popular in Darwin's day and of special interest to him. The most exotic kinds of pigeons resulted from such human tinkering with tufts of feathers in the oddest of places and a range of exotic behavior as well. The English were also fascinated in breeding exotic roses and entering their creations in contests. Plants cooperate with human hybridization attempts more readily than do animals, but both can be altered.

We wonder how humans ever got started in domestication in the first place. The dog, or wolf, as it was back then, is most likely the first wild animal to have been domesticated. Some estimates of when that first happened go back to twenty-five thousand years ago or even further. Maybe it began with a human child wanting to keep a lost baby wolf as a pet. No matter how it got started, both dogs and humans gained from the new partnership. The first dogs stood to gain a steady supply of food by being around human hunters. The humans gained a companion with keener senses than they had. The dog has far superior hearing and that made it a natural guardian. It is interesting to note that dogs bark, whereas wolves do not. The dog has an extraordinary sense of smell and that made it a superior tracker. Dogs are social animals and that made them good team players. The key to creating dogs from wolves was to eliminate the natural viciousness in these camp followers by either killing the more vicious ones or not allowing them to breed. Rewarding good behavior with food may have been important too. Breeding dogs with customized traits continued through the generations. There are over different four hundred breeds of dogs today.

With the image of friendly dogs in my mind, I am reminded of the Silver Fox experiment and what it can tell us about domestication. Dimitry Belyaev studied silver foxes over a fifty-year period in Russia. Now, foxes are naturally wary of humans and retreat from them. If contact is forced, they may bite the offending human. With this in mind, Dimitry intentionally selected the friendliest 20 percent of fox pups for a study group. They would be allowed to breed with each other at adulthood but not with the less friendly foxes. He also kept a control group of normal foxes going for comparison. The experiment ran for forty generations. It turns out that the special group became radically different from the control group. The sought-for friendly behavior was achieved in spades. The friendly foxes really wanted to hang out with humans, they whimpered for attention, and they even licked their caretakers. However, it was how these foxes changed physically that was so amazing! They developed floppy ears, short and curly tails, new fur colorations, and different skulls and teeth and lost their musky smell. So, the power of domestication can be an awesome thing.

There is a scientific aspect to domestication too. Domesticated animals are genetically different from their wild counterparts. Just think of how radically those silver foxes had changed. Their DNA had changed from the selection process. Although Darwin formulated the concept of artificial selection, he never understood how inheritance of traits worked. So, we have the advantage over him in understanding that domestication is a change in the genetic code of the animal over generations. That is why dogs of today love being near humans, whereas wolves behave in a directly opposite manner. Think of the milk cow with its bony appearance and huge udder. It has been altered into a milk-producing machine. It has been altered so drastically that it can no longer survive in the wild.

For those readers, who want to delve deeper into the domestication activities of the pioneer farmers during the Agricultural revolution, I recommend Jared Diamond's acclaimed book, ***Guns, Germs, and Steel***. You will come away with a deep appreciation of what these first farmers experienced and accomplished. The crops, that we take for granted today, were the result of masterful hybrid experiments with specific objections in mind.

Human Evolution during the Agricultural Revolution

Infectious Diseases and Malnutrition

The lifestyles for farmers were different than they were for herders. Farmers increasingly needed to stay put on their farms to protect their crops and livestock from nature and from thieves. Herders, on the other hand, needed to be nomadic. Once a pasture was exhausted, the herders needed to move their herds to fresh pastures. The concepts of germs and proper sanitation were not understood during the agricultural revolution with the result that new diseases developed and rapidly spread through the communities. Rodents attracted to the grain, and food waste spread diseases like typhus and bubonic plague. However, diseases also developed from proximity to farm animals. Farmers tended to live very close to these animals, and the opportunity for diseases jumping to humans was rife. Most of the childhood diseases, which we could name, evolved during this period. The difference is that these early farmers had no immunity to them yet, so the diseases were often lethal. When I was a child growing up in Illinois, I had mumps, chickenpox, measles, whooping cough, and other diseases without worry of dying from them. I can thank my distant European ancestors for my immunity to them. Generations of them suffered in order to evolve that immunity. The native peoples of the American continents were not so fortunate. They had no domestic animals in their agricultural revolution and consequently no need to evolve immunity to such diseases. So when the Europeans invaded the Americas in the sixteenth and seventeenth centuries, the contagious diseases they brought killed the vast majority of those defenseless Native Americans.

Contagious diseases are not the only health problems caused by the agricultural revolution. Vitamin deficiency and malnutrition were serious problems too. Relying on cereals excessively meant that these people were no longer getting the varied diet that they did as hunter-gatherers. For example, a lack of thiamine in their diet led to beriberi with damage to heart, circulation, nerves, or muscles. Eating beans, vegetables, or meat could have corrected that problem. Other vitamin-deficiency diseases included pellagra from lack of vitamin B_3, rickets from lack of vitamin D, and scurvy from lack of vitamin C.

The mobile herders were generally healthier than the stationary farmers because they didn't live in their own filth. Yet, they gained even a greater advantage over the farmers when they evolved lactose tolerance and could drink milk from their animals without painful indigestion. Here is the situation they confronted: human babies and infants can digest milk until the age when they are naturally weaned, and then they lose that ability. They become lactose intolerant. Now out of hundreds of herders, perhaps one or two individuals were genetically lactose tolerant as adults. That seemingly small advantage had such a powerful survival value that over the generations, it grew and grew in frequency. Milk is a very nutritious food, and it contains vitamin D. Today, over 70 percent of the population of northern Europe is lactose tolerant due to their reliance on milk for sustenance.

The bottom line is that the agricultural revolution in Eurasia brought severe disease and malnutrition, which affected generations for hundreds of years. It has been determined that the average height of people from this period is significantly shorter than that of the hunter-gatherers who preceded them. On the other hand, agriculture brought stable food supplies and accelerated population growth. The advantages must have outweighed the disadvantages because agriculture steadily displaced hunting and gathering as a way of life throughout the world.

Personality Changes

Authors Cochran and Harpending propose that the agricultural revolution not only changed humans in terms of adapting to a radically different diet, but also changed humans in terms of personality. Humans had to adjust from an egalitarian society to a stratified society, where ownership of lands and property was important, and where some people had life or death power over others. For example, the concept of delayed gratification was unknown to the hunter-gatherer. If you killed a caribou, you may as well share it with others. The meat won't keep for long anyway. However, as a farmer, you must be selfish and not consume your seed crop even if you are low on food now. You certainly must not give it charitably away to hungry neighbors, or there will be nothing to plant in the spring. People who could adapt to this new way of behaving, survived and raised their families successfully. Those who could not adjust, vanished from the gene pool. Humans were learning the art of domesticating plants and animals. In the process of doing that, they also domesticated themselves.

Religion and the Rise of City-States

Mesopotamia was an area within the Fertile Crescent, which is known as the place where civilization began. Lying between Tigris and Euphrates Rivers, the Mesopotamia of one hundred centuries ago had rich fertile soil for growing crops. Why is that important? Because a stable food supply is the first essential step in building a city-state. We find that this is true for all the civilizations getting established throughout the world. A sense of community is another element needed, and it is here that the role of religion is so evident. Most major religions of today are defined by an all-powerful god, but for thousands of years, the custom was to believe in multiple gods. And so it was in ancient

Mesopotamia. There was Anu, god of the sky; and his son, Enki, god of the earth. Anu had another son, Enlil, who was god of the wind. The Babylonians were one of the many peoples of Mesopotamia, and their main god was Marduk, son of Enki. The list of gods goes on and on.

The various city-states shared gods or may have had their own unique god. In any case, they needed to visualize their gods to feel a connection. So, they made a statute image of their god and believed that once constructed and placed in its temple, that statute would contain the real god and be empowered. To this end, they expended thousands of manhours to build their temple at city center. Over the centuries, these temples became more elaborate and multistoried. Such temples were called "ziggurats" and were built from mud bricks. These bricks were the standard building material in Mesopotamia, whereas limestone was the more common material in Egypt and Greece.

The rise of civilization in Mesopotamia did not occur in isolation of the rest of the world. We should understand that trading had been evolving from thousands of years previous into a bigger, interregional, lucrative business that helped finance the growth of these city-states. Moreover, traders returning home carried stories of the wonders they had seen in foreign lands. This caused a cross-fertilization of innovative ideas between regions. The Nile delta was another fertile region, where agriculture led to rapid advance to city-states and kingdoms. The first Egyptian dynasty was founded in 3100 BCE, and writing was developed there, concurrent with the development of writing in Mesopotamia. The Mediterranean Sea was not a barrier either. Seafaring traders connected the Greek Islands into the civilized world. We know this because the archeological finds in ancient Crete are quite impressive. Trade routes even connected Europe to India and China, where agriculture and civilization were sprouting as well. More mysterious is the advance of civilizations unknown to the western world. For example, the Mayan civilization was emerging as early as 2600 BCE in Mexico. These Mayans went on to build colossal structures, develop writing and astronomy.

Innovation on Steroids

The seed of great accomplishment must have existed in our hunter-gatherer ancestors all along, but it took the combination of an agricultural economy, a stable government, and diversity of human endeavor for that seed to take root. During the great leap forward, rapid advancement in stone toolmaking, antler and ivory implements, and sewn clothing took place. The agricultural revolution brought a redirection of stone toolmaking to support farming needs. This new Stone Age culture is known as the Neolithic period. Grinding stone tools to a sharp edge was one innovation, but there are numerous others.

Perhaps, the invention of pottery was the most significant of the Neolithic advancements. It began with the hand shaping of a clay snake into a pitcher or other vessel and baking the object to harden it. However, when the potter's wheel was invented in 3500 BCE, it accelerated the shaping process, produced symmetric products, and increased output measurably. Archeologists are beholden to the invention of pottery and its widespread use in Neolithic times. Those pottery shards discovered in excavations are valuable because they provide important clues to the identity of prehistoric cultures and the way they lived. Clay and mud had yet another use, that of forming bricks. This invention revolutionized the kind of dwellings in which people lived. The idea of staying in one place was not unheard of in hunter-gatherer times. Sometimes they built huts with sticks and pelts and stayed put for months. However, they were unwilling to expend the energy necessary to build a brick house. It took the Agricultural revolution to change that mind-set.

The domestication of animals had advantages beyond a ready source of meat. Human labor could be done instead by much stronger animals. For example, when the plow was invented in 4500 BCE, oxen were recruited to pull those plows. Horses were used to pull carts and chariots. Dogs were trained to herd grazing animals and protect them from predators. Sheep did none of these things, but they did part with their wooly coats for us. We used that wool to make clothing. As soon as wool spinning was invented around 4000 BCE, a textile industry sprung up providing the region with items to trade.

The first use of the wheel for transportation occurred in Mesopotamia about 3200 BCE. One wonders what took so long. With our advantage of hindsight, the wheel and its usefulness seem so obvious. Yet, it never independently materialized in the Americas. The Native Americans first learned of the wheel when they saw their European invaders using horse-drawn wheeled carts. The first Mesopotamian wheels were clunky! Two or more boards were bolted together and cut to a circular shape. An axle was attached through the centers of the wheels. Useful inventions seem to be quickly adapted to warfare and so it was with wheeled chariots. Horse-drawn chariots became lethal weapons of warfare soon after the wheel appeared. One person could hurl spears at the enemy, while another person drove the chariot. Spooked wheels were an improvement over solid wooden wheels, but it took centuries for them to appear. The first spooked wheels appeared on Egyptian chariots around 2000 BCE.

Writing may be the greatest invention that humans ever made. Without it, we would not know of these wondrous historical times, but more importantly, humans wouldn't be able to transfer knowledge across the generations and use that information to build and improve on. Sumer was a major civilization in southern Mesopotamia, and it is credited with inventing writing around 3100 BCE. They used a method called "cuneiform." It consisted of making impressions in a clay tablet using a stylus. Writing was invented as a means for keeping track of agricultural production, but it soon found application in a variety of other fields. Of course, general communication between humans soon became its imminent purpose. It seems the world was ready for writing because it sprung up around the world in a variety of formats. The Egyptians developed writing at about the same time as the Mesopotamians, but they used a different method called hieroglyphics or picture art. The early Greeks had writing, and even the far-off Mayans in Mexico developed a form of writing.

The Neolithic period ended when metals were found that could do the job better than stone tools. Copper was the first of the metals to replace stone tools, and we refer to this period as the "Copper Age." Now, copper was initially obtained in the metallic state from meteorite outcroppings. It had special properties, which made it valuable. For one, it wasn't brittle and subject to fracturing. For another, it was malleable. Those copper nuggets could easily be hammered and formed into useful shapes, like bowls. When was copper first used? Its discovery probably dates back to the dawn of the Agricultural Revolution. We know that the metal was in use around 8700 BCE. A copper pendant was found in northern Iraq, traceable to that early date.

Now, the popularity of copper really increased when our ancestors learned that it could be melted and cast in a mold. Casting is a process where one can duplicate an elaborately prepared object quickly and cheaply. We know that copper casting was being done in Turkey around 6400 BCE and that it had spread to Egypt by 4500 BCE. Innovation was rampant as these early civilizations arose in the Fertile Crescent. We know that someone solved the problem of extracting copper from its ore by 3800 BCE because Egyptian records describe the mining. The Copper Age was born as metallic copper became readily available. The Copper Age (3500 to 2300 BCE) was a paradigm shift, where copper replaced stone tools in many applications. However, it was mainly a regional phenomenon, restricted to the Eastern Mediterranean.

Copper has one serious flaw though; it is too soft. Our innovative ancestors didn't let that flaw stop them from staying with metal to help them solve problems. They found a way to make copper harder. After trial-and-error attempts, they discovered that adding tin or arsenic to molten copper solved the problem by making copper harder. The new metal is called bronze, and customers lined up to use it. So much so that the period is named the "Bronze Age" (3300 to 1200 BCE). This harder metal could hold an edge and that fact revolutionized warfare. For example, swords, arrowheads, and spear points could be mass-produced by using a casting operation.

The next leap forward in weaponry came with the processing of iron ore to produce iron implements and eventually steel ones. Our history is filled with stories of medieval battles and steel weapons. A by-product of copper smelting was glass beads. Someone may have wondered if it had a use and experimented with it. By 1600 BCE, the technology of produced glass from silica was discovered, and the glass industry got its start.

The Monument Builders

One of the wondrous things we associate with Egyptians is the marvelous pyramids still standing after thousands of years. The realization that people with ancient tools cut huge blocks of limestone out of quarries, hauled them miles from quarry to construction site, and raised them into position in the pyramid is amazing. Equally astounding is the large number of these massive monuments. There may be as many as 138 pyramids in Egypt. How long ago was that? The earliest pyramids were built by the third dynasty around 2630 BCE. If size impresses you, then the pyramids at Giza are worth checking out. They are among the largest constructions in the world, and they are in the outskirts of Cairo. The Pyramid of Khufu is the largest Egyptian pyramid and one of the wonders of the ancient world. It has been measured with modern instruments and deemed to be a perfect square with ninety-degree corners. It contains two million limestone blocks. The king's chamber is made from red granite, which was quarried hundreds of miles away and is much harder to work on than limestone.

Now, these pyramids tell us something about the power of religious belief and its influence over human behavior. The pharaohs of Egypt were regarded as gods, and the pyramids were their burial chambers from which to ascend into heaven. The pyramids not only housed their bodies but their needs in the afterlife, including food, wine, furniture, chariots, and the like. Whatever inspired the ancient Egyptians to expend vast resources to building monuments, they were not alone in the world.

Figure 9.2 The Great Pyramids of Giza
By Ricardo Liberato ([1]) [CC BY-SA 2.0 (http://creativecommons.org/licenses/by-sa/2.0)], via Wikimedia Commons.

In the following table, I have listed a few of the many megalithic monuments built during these early days of civilization. I wanted the reader to understand that something extraordinary is going on here. The fact that our ancestors after being hunter-gatherers for two to three million years, where an egalitarian, nomadic lifestyle was dominant, could so quickly transform into a kind of society where huge sacrifices in time and labor are being made for a symbolic purpose is mind boggling. Part of the transformation story has to do with the rise of a ruling class, but the significance of religious belief must be important too. We know that slaves were captured in the ongoing wars between the city-states, and these slaves were forced to work on these ambitious projects. However, there were also many others who worked enthusiastically on these projects as skilled craftsmen and artisans. There is bountiful evidence of ingenuity and precision in these structures that is unexpected for a Stone Age people.

Table 9.1 Examples of Monument Building in the Ancient World

Region	Initial Period	Monument
Turkey	10,000 BCE	Gobekli Tepe
England	3000 BCE	Stonehenge
Egypt	2500 BCE	The Great Sphinx at Giza
Iraq	2000 BCE	The Ziggurat of Ur
Greece	447 BCE	The Parthenon at Athens
China	210 BCE	Mount Li -Emperor's Tomb
Peru	1436 AD	The Sky City of Machu Picchu
Easter Island	1000 AD	Statues of Aku-Aku

Gobekli Tepe is the anomaly in this list due to its great age. Traditional thinking tells us that the agricultural revolution created the conditions under which great stone structures could be erected, and yet here is a great stone structure under construction before the agricultural revolution began. This site built upon a mountain ridge in southeastern Turkey is relatively new to the world in that its excavation began in 1995. The leader of the excavation, Klaus Schmidt, passed away in 2014 but while alive had expressed the opposite view of the traditional. He felt that temple building came first and that agriculture was invented to feed the workers on such projects. If Schmidt is correct, then the intellectual level of these hunter-gatherers was greater than we would have guessed because the planning involved an engineering knowledge and the shaping and moving of multi-ton stones was a challenging prospect in primitive times.

Gobekli Tepe is unique in having two tall T-shaped pillars in the middle of the enclosures. They weigh up to sixteen tons and are carved with arms, hands, and loincloths. The site is built upon mounds of animal bones and stone vessels, indicating that this site had long been a gathering point for feasts, social interaction, and worship. In the overall area, over two hundred pillars in about twenty circles have been identified. The pillars are up to twenty feet tall.

Stonehenge is a prehistoric monument in Wiltshire, England, that was constructed between 3000 and 2000 BCE. The site is noted for its enormous sarsen stones set in a circular pattern and its special bluestones, which were quarried from far away and transported to the site. While Stonehenge is predominately a religious structure, it also had some astronomical significance. The time of the summer solstice is captured when certain features line up with the morning sun.

The Great Sphinx is as famous as the great pyramids that dwarf it at Giza. The Sphinx is mysterious in that it has the body of a lion and the head of a man. The powerful body of a lion was associated symbolically with the also powerful Sun God. The monument stands sixty-six feet tall, and the body is 236 feet long. Its construction is credited to King Khafre, who reigned from 2520 to 2494 BCE. It is very likely that the face of the Sphinx is that of King Khafre. The Sphinx was constructed of stone at the site. In fact, a trench was excavated around the body of the Sphinx and the blocks used for the upper portions.

The Ziggurat of Ur is an example of religious temple building in ancient Mesopotamia. The people building it are the Sumerians, who are also famous for inventing writing. The common building material of Mesopotamia is mud brick and that is what was used to build this temple. The base is 205 × 141 feet, and the structure consists of three levels. The walls of the first level are 36 feet tall and slope inward. The second level is 18.7 feet tall, and the third level is 9.5 feet tall. One long sloping stairway traverses the three levels and enters the shrine on top. Two other stairways are at right angles to the long stairway, and they meet the long stairway at the top of the first level.

The Parthenon at Athens is an example of masterful temple building by the ancient Greeks. This temple held the statue of the goddess Athena, from whom Athens got its name. This is a structure, which I personally witnessed twice in my life. The first time was when I was in the US Navy on a Mediterranean cruise. The second was more recently when my wife and I took a Princess Cruise tour of the Greek Islands. As we visited the island museums, it seemed as if we had been transported thousands of years in the past, and yet the quality of carved marble statutes was exquisite. As for the Parthenon, it lies elevated above the city of Athens on a high mound. The complex of ancient buildings on that mound is called the Acropolis. The Parthenon that exists today is a reconstruction of the temple after the Persians plundered the temple in 480 BCE. The temple is 228 by 110 feet in size and is most recognized for its Doric columns. Huge blocks of limestone were hauled from miles away, brought up the hill, and in some cases, raised to rest upon these Doric columns of which there are eight on the ends and seventeen on the flanks. This temple is another example of the powerful influence of religion on the people of the time.

Mount Li is the tomb of China's first emperor. He conquered the separate provinces and united them into one empire. As for his tomb, an enormous expenditure of manpower was devoted to building it; over seven hundred thousand men were conscripted to the effort. Most of us know of it due to the life-size terra-cotta army that was found buried underground. Each soldier is individualized and beautifully sculpted. The penalty for imperfection was the artisan's death. However, the eight-thousand-strong army was just a small part of the effort to duplicate the known world in the burial site. The entire burial site occupies one square mile. Besides the terra-cotta army, two exquisite bronze chariots each with four horses and driver were discovered. The imperial stables were also duplicated. The tomb itself has not been opened, but it is expected to consist of numerous rooms with jade carvings and other treasures based on similar excavations. It seems to be a common trend during these early times for great leaders to be regarded as gods and great efforts are made to honor them in death.

The Sky City of Machu Picchu lies on a mountain ridge nearly eight thousand feet above sea level. This fascinating city contains many mysteries, but scientists believe it was built as an estate for the Inca

Emperor Pachacuti (1438–1472). The Incas were essentially a Stone Age culture, and the Spanish invaders had yet to appear on the continent. The workmanship and intricate planning of Machu Picchu make it one of the new Seven Wonders of the World. It seems to have been engineered to survive for ages. It is situated in a seismic area where earthquakes and landslides are common, and yet this city is still intact. That may be due to how it was constructed and that is from irregular-shaped blocks of granite. The blocks fit so well together that one is unable to fit a knife blade between them. Like many Stone Age people, the Incas worshiped the Sun and built a temple within the city to honor the Sun God. The Incas were thought not to have developed a written language, but that conclusion is being reconsidered. Knotted strings may hold the key to recovering their ancient history.

The Statues of Aku-Aku were made famous in the 1957-book *Aku-Aku: The Secret of Easter Island* by Thor Heyerdahl. These sixty-foot-tall statutes are heads and upper bodies carved in large blocks of stone. Archaeologists have documented as many as 887 of these massive statutes, but there may be others. The statutes were carved between 1100 and 1680 AD. These Polynesian artisans were unaware of western technology and lived in an essentially Stone Age culture. So it is a mystery as to what motivated these islanders to expend such great resources in the carving and moving of multi-ton statutes to strategic points on the island.

The Rise of Empires

As the numerous city-states developed in Mesopotamia, competition and hostility between them naturally occurred. Rulers emerged of varying skill in governing and defending their territory. These rulers had power unlike anything experienced by humankind before. They maintained order and made laws. They were the supreme spiritual leaders and sometimes were regarded as minor gods. They directed the efforts to build ever-grander temples of worship. They were also commanders of their armies. Eventually, some rulers felt they were powerful enough to rule two or more city-states. This gave birth to the age of emperors and empires. For example, the name Sargon the Great is still remembered today. He was king of Akkad, an area south of today's Baghdad. He conquered the Sumerians to the south and went on to occupy most of Mesopotamia. His descendants ruled the empire for a century after his death.

Jumping ahead about two millennia, we come across the name of the first world conqueror (i.e., the known world), Alexander the Great (336–323 BCE). He inherited the throne of Macedonia, one of the Greek colonies. His father, Phillip II, was away in military campaigns during Alexander's youth but provided his son with the best in tutors. One of those tutors was Aristotle. Alexander proved to be an even better commander than his father. He united the Greek city-states and conquered Persia (i.e., Iran), Babylon, and much of Asia.

Thanks to the invention of writing, we have a detailed history of the world from Sargon's time and forward. The story of growing empires was repeated again and again in other parts of the world. Even in the American continents, the Aztec, Mayan, and Incan empires followed the pattern begun in Mesopotamia. They never developed metals like bronze and iron, domesticated animals like the Eurasians, or even invented the wheel, but they did build advanced cities with large temples, roadways, irrigation, and developed corn and potatoes. From this point in history up until today is too much to cover in any detail. Consequently, I will finish the chapter with some general observations.

Between the Dawn of Civilization and Today

Population and Culture

Agriculture is at the heart of human progress. It started in areas having wild plants and animals, which were amenable to domestication. The new lifestyle began in the Fertile Crescent and spread to the Atlantic

Ocean in the west and to the Pacific Ocean in the east. Wherever agriculture took root, the resulting abundance of food allowed populations to grow and prosper. Bigger populations have the potential to make progress faster than small populations. This is because problems get solved faster when more minds are working on them. That is one of the benefits of a growing population, and the Agricultural Revolution caused a paradigm change in human life. Our planet has been transformed, and we see it in today's world. The wild and natural world of our hunter-gatherer ancestors is virtually gone. In its place, we observe a world covered with wheat fields, cornfields, rice paddies, and other human industries. Where bison once grazed by the millions on the American plains, cattle ranches, sheep ranches, pig farms, chicken and turkey farms exist instead. Wildlife is in trouble: for if an animal is not fit for domestication, it may well be on a path to extinction. Even our great ape cousins are on the endangered list.

In a sense, humans have been undergoing domestication too. Agriculture led to permanent populations, which led to villages, then cities, and then nations. A different kind of social structure was required to coordinate these larger groups. A ruling class was needed for one. Civilization brought a differentiation in human roles that didn't exist during our long hunter-gatherer period. Property ownership and the accumulation of wealth were new to humankind until the agricultural revolution. The nomadic hunter-gatherers could only possess what they could carry with them, but civilization changed all that, and wealth and power became agents of change to humanity. Humans were more independent and probably more dangerous to cross in hunter-gatherer times. The division into different classes of people changed those personalities. If your family's livelihood depends on obeying and being subservient to your superior, then you had better adapt. Some fifty thousand generations of humans have existed since the dawn of civilization, and each generation has become less independent and less violent than the one before. There is evidence that we have become progressively more intelligent too. Now the world population exceeds seven billion people. Humanity is still changing at an observable rate.

Religion

The early stages of a spiritual awareness were evident in the Cro-Magnon times as far back as fifty thousand years ago. Ritual burial is perhaps our best proof of that awareness. When graves were decorated with thousands of handmade ivory beads, we can be pretty sure that our ancient relatives believed that the dead would live on in some spirit world. Sometime in this period, a type of spiritual leader arose. These shamans provided answers that the tribal leaders could not provide. Why is my child dying? Can anyone save him? What happens to us when we die? It is also likely that the shaman were the custodians of medical knowledge. Their success in saving lives now and then gave them credibility. As populations grew during the agricultural revolution and a stratified culture developed along with the rise of villages and small cities. We have seen that religion played a major role in the rise of city-states. Ever more elaborate temples were being built in the center of their cities. Polytheistic religions ruled the day and still persist in some regions of the world. How important were these religions to the rise of civilization? Apparently, they were vitally important. After all, religions can bind societies closer together and give them a more unified purpose. On the other hand, they may occasionally prevent societies from doing what is vital for their survival. Religions succeeded or failed based on their influence in usefully modifying their believers.

Whereas the Mesopotamians, Egyptians, and others worshipped the sun, the moon, the wind, and other natural phenomena, a form of these gods with personalities soon emerged. For example, the Greek gods had Eros who was the god of love, procreation, and sexual desire. Other Greek gods included Aphrodite, goddess of love and beauty; and Athena, goddess of wisdom, poetry, art, and war strategy. Moreover, these gods and goddesses interacted with mortals on a regular basis. The Greek gods were adopted by the Romans and given different names. Although we refer to these religions as mythology today, they were just as real to the ancient Greeks and Romans as our religions are to us today. Meanwhile, the Israelites were occupied by the Romans and undergoing a radical change in their religious beliefs as Jesus Christ taught a religion of tolerance and forgiveness. The Vikings, who were closer in time to us, believed in polytheistic gods too. They resisted having Christianity imposed on them for centuries but finally submitted.

Today, there are major religions having a billion or more members. Two of them, Christianity and Islam, seem to be in a competition in the world. Both of them at one time had zero tolerance for dissenters. The Holy Roman Church was the dominant Christian faith for centuries. However, the protestant movement broke that hold, and today, hundreds of different Christian faiths now exist. The Muslims have divided into Shite and Sunni, and they seem much the same to western eyes. However, considering the hostility between them, they must see it differently. The United States, which is predominately a Christian nation, has a constitutional provision fostering freedom of individual choice in religion. Western Europe is similar to the United States in this regard. However, the predominately Muslim nations see it differently and strongly discourage freedom of religious choice. This dispute has been the focus of the world's attention for the last few decades at least and doesn't seem to be likely to be resolved any time soon.

Conflict

Civilization brought many benefits to humankind, but at times it also brought agony in the form of warfare. We can't know much about the conflicts that arose during hunter-gatherer times, but they were minor clashes compared to scale of warfare conducted by civilized societies.

We already discussed how warfare between the early city-states arose. The Bronze Age led to mass production of armor, spears, and swords. The horse-drawn chariots were an awesome weapon of their time. It seems that metallurgy has a strong relationship to success on the battlefield. Iron swords made bronze swords obsolete. The development of steel was another big step forward in weapons. The Chinese developed gunpowder, but it soon was used in warfare around the world. These early battles were intense by today's standards. The goal seemed to be to totally destroy the opposing army. Even those who surrendered were executed or enslaved. Today, the rules of war are more humane, but the weapons of war more lethal. Machine guns, napalm, landmines, and the like present grim images of where we have arrived.

World War II was underway when I was a child. I have seen movies and read countless books about that period and the main characters in it. Some of these seem heroic and exciting, but a soberer judgment is that it was a horrendous experience for tens of millions who suffered and died in it. Many of those deaths were not from weapons, but from starvation or disease. Living in the United States was fortunate for us because the war never made it to us here, but in many other parts of the world, there was no escape possible. Hopelessness is worst situation we humans can imagine.

World War II did one other world-changing thing; it introduced nuclear weapons onto the world stage. The postwar period became a worldwide alignment of nations influenced by either the United States and its democratic partners or the Soviet Union and its allies. An arms race of more and more lethal nuclear weapons threatened humanity itself. I happened to work in that industry in the early sixties and will never forget working seven days per week for an entire year as the nation raced to build up our nuclear weapons stockpile. I begged for a day off from work to no avail during that period. The Cuba Missile Crisis in October of 1962 could have ended in a nuclear exchange between these super powers, had a different president or soviet leader had been in power. President Kennedy's advisors were urging him to make a first strike as the tensions and bluster escalated. The aggressive nature of humans is in our genes. Evolution made us that way in a totally different kind of world. Our hormones race through our bodies to rapidly prepare us for a fight-or-flight scenario. Our social nature is to cry out for confrontation or revenge. However, our best interests are served by remaining calm and rational. We humans are in a battle against our natural impulses. Our technical power has gotten far ahead of our ability to subdue our emotions.

Science and Technology

Innovation was the theme of the Great Leap Forward, which began some fifty thousand years ago. However, it really gained momentum during the Agricultural Revolution. One of the big steps forward came with the invention of writing some five thousand years ago in Sumer. It was first used to help keep track of trades, as trades became larger and complex with time. How many bushels of grain would you take for your five oxen? Impressions in clay tablets sufficed for early commerce, but written communications gained complexity. Pictographs gave way to alphabets, and before long writing has used for a myriad of uses besides commerce. What we know of ancient civilizations come from deciphering their written records. Writing has to be one of the most transformative inventions of humankind. It allowed communication at a distance, communication between generations, and for the first time, we had a way of recording history and learning from those before us.

We think of the western world as the technological frontier of humankind. Indeed, it is. However, it might have been different. The Chinese were far ahead of the west at one time. However, their government of that time decided to apply the brakes to innovation. Jared Diamond, in *Guns, Germs, and Steel*" argues that their strong centralized government made that cultural decision happen. Power in the western world was more fragmented, and so a similar mandate in the west could not be enforced. The lesson is that fragmented power is more conducive to technological progress that centralized power. The comparative geography of China versus Western Europe is an important factor because barriers in the latter made fragmentation more likely.

Beginning in the sixteenth or seventeenth century, science and technology began to exert increasing influence over human affairs. Religion had ruled supremely until this time, and its struggle to remain dominant was fierce. Science had one big advantage, it provided results. Moreover, that innovative spirit that materialized fifty thousand years ago still was strong, and it didn't want to be stifled. The Catholic Church was a powerful force in the seventeenth century, and it had a historic clash with science. Galileo Galilei had used his telescope to observe the four large moons of Jupiter. He recorded his observations and became convinced that they made regular circuits around the planet. He also observed the orbit of the planet Venus over time. These observations convinced Galileo that Nicolas Copernicus was correct in 1543 when he proclaimed that the earth orbits the sun and not the other way around as commonly believed. The church reacted forcefully to this challenge. In 1616 the inquisition declared heliocentrism to be heresy. Galileo voiced his views despite the threat and was tried for heresy in 1633. Perhaps the next big battle was in 1859, when Charles Darwin published *On the Origin of Species* where he described his theory of evolution. The church position was that species were immutable. God created perfect plants and animals, and they remained in their perfect form. Darwin believed the opposite. Living things adapt to best fit their environment. Darwin didn't face a prison term as did Galileo, but the battle isn't over even in the twenty-first century.

Amazing scientific and technological progress has been made since then. Steam replaced sailing ships and powered locomotive trains in making the world smaller. The Model T saved the world from wading in horse droppings. Hybrid autos and electric cars are in the process of replacing the gasoline cars. Gaslights and candles have been replaced with electric lighting. Coal and oil have fueled those electric power plants, but solar power and wind power are coming on strong. My parents listened to their favorite programs on the radio when I was young; black-and-white TV replaced that activity and color soon followed. Now, we watch programs on our portable devices. Those devices also know everything, and we consult them often when our memory fails us. Medicine was crude a few centuries back, and we didn't know that germs caused disease. Sanitation was a huge step forward, but these days we know the secret of life and are manipulating our genomes. The rate of scientific progress is so awesome that no one grasps the entire picture.

Ten

The World Ahead

The Chapter in a Nutshell

The world population of humans has been growing exponentially and is now over seven billion people. The complexity of life due to technological advances has been increasing at an astounding rate too. To cope with today's complexity requires mental competence. To thrive in today's societies requires superior intelligence. Selection processes are shaping the modern brain, but many suffer from the stress of competing in this rapidly changing world.

Modern life eliminates the major fears our ancestors faced such as starvation, exposure, predators, and medical emergencies. That makes us wonder if humans are still evolving or if modern technology has erased all culling mechanisms of natural selection. Two scientists, whom we will discuss, believe that humans are evolving at an astounding rate, mainly due to the enormous population size.

The discovery of DNA's structure was a landmark event. It led to man's ability to tinker with the genetic code and modify plants and animals to suit human needs. Now, the ability to tinker with human genetics is in our power. Already, modern miracles are giving hope or even improvement to previously unsolvable problems like blindness, deafness, mental illness, and others. The future for people born with defective genes is not hopeless. A process called gene therapy holds the promise of introducing healthy genes into such patients. It has already helped autoimmune children and some cancer patients.

Advancements are also being made in attaching mechanical limbs or exoskeletons to paralyzed patients and patients with missing limbs. The cyborgs of science fiction are becoming more feasible every day. Ideally, the wearer of these devices will be able to move them with thought control and get sensory feedback in order to employ fine motor control. For example, to have the capability of picking up an egg without breaking it. Research into restoring mental control of paralyzed limbs by rewiring nerves to function again is making progress.

We Have Come a Long Way as a Species

The Homo lineage story includes an over two-million-year evolutionary transition from some yet undesignated bipedal ape species, that survived on vegetation, to human hunter-gatherers, who became reliant on having meat in their diet. The benefits of that meat diet appear to have been enormous. Predominately, we see there was a nearly three-fold increase in brain size offset to some degree by a much smaller gut. Larger brains must have been vitally important to our ancestor's survival because maintaining such a large brain is costly in calories and costly in terms of a large investment in infant care. Whatever the benefits of a larger brain during that transitionary period, we can certainly see the benefits of a better thinking and more socialized mind in our world of today. Humankind has bent the world to our will and maximized our safety and survival.

The pace of human creativity proceeded very slowly until about fifty thousand years ago, when the Great Leap Forward propelled it to very noticeable level. For the first time, humans moved beyond a life of merely surviving to a spiritual awakening, illustrated by decorated tools and weapons, expressive cave art, jewelry, musical instruments, and ritualistic burial. Moreover, the survival ability of these pioneers was so advanced that they were able to migrate into new lands with new and unexpected challenges. We know they handled the challenges well because their populations blossomed. Eventually, our ancestors occupied all the continents except Antarctica.

The hunter-gatherer lifestyle had defined us for two million years, but the agricultural revolution, which began twelve thousand years ago, was going to change all that. Plants and animals were domesticated and tailored to serve humankind. Problems of disease, malnutrition, and societal change accompanied that revolution, and humans evolved new mechanisms to cope with the changes. Villages appeared for the first time but soon grew into cities, nations, and empires. These massive cultural changes transformed humankind forever. The individual, who previously enjoyed autonomy, was becoming more and more a pawn of the establishment. Like their own farm animals, humans were becoming domesticated too.

In the battle between man and nature, nature occasionally wins a few rounds too. Civilization was well advanced in the fourteenth century, and yet the Black Plague struck with devastating effect. Europe lost up to two-thirds of its citizens. Bubonic plague is still with us today, but the advances in sanitation and medicine reduce it to only a minor threat.

Is Natural Selection No Longer Operative for Humans?

The Human World Is Less Threatening Now

Some scientists believe that humans stopped evolving since the advent of civilization, sanitation, modern medicine, mechanized farming, and the welfare state. It would seem that the culling mechanisms of disease, injury, malnutrition, predators, and other lethal agents have been virtually eliminated from society. We witness how well Darwin's principle of natural selection operates in nature, but we question whether natural selection still operates in the human world. For a long time now, humans have cooperated to prevent those dangers from threatening us. For example, starvation is not the constant threat it was in ancient times. Advanced agricultural techniques produce food in excess of our needs. Disease has not been eliminated but is nothing like the terrible threat it has been in the past. The discovery of germs as a cause of disease and the institution of sanitation measures made a huge difference to prevent the spread of disease. Medical science is making enormous strides in curing or at least alleviating injury and disease. Few of us live on the frontier anymore, so we live in a safer world and look forward to living to retirement age and beyond. Some scientists have declared that humans are no longer evolving because the culling mechanisms of natural selection no longer operate in our lives.

The "Human Evolution Has Ended" crowd sometimes fail to recognize that natural selection is not the only mechanism advancing human evolution, yet they argue as if it was. However, there are other mechanisms, which influence the course of our evolution. They include the founder's effect, genetic drift, and gene flow. Moreover, these mechanisms are very significant to our evolution too. Let's examine each of these mechanisms.

Founder's effect is when a group splits off and becomes isolated from the general population. If the group is not a typical genetic representation of the general population, but in fact is quite untypical or becomes untypical due to members dying, then their offspring can evolve in ways even more untypical of the original population. In other words, these people lack the full diversity of the general population and react to new stresses differently because they are unable to react as the general population would have.

Genetic drift has to do with the fact that the same gene can exist in two different forms, that is, alleles. Each of us have pairs of genes, each contributed by one of our parents, but only one of them is selected when our bodies manufacture a sperm or an egg. It well might be that these sperms and eggs contain the rarer alleles. So, it is a dice roll as to which alleles make their way into the embryo. When a population is small, genetic drift due to random selection of competing alleles is a possible result. Again, like for the founder's effect, the genetic composition of the small group becomes untypical of the original population and has to react to future stresses differently than the general population would.

Gene flow has to do with change in offspring traits due to mating with very different kinds of partners compared to mating only with locals. In the past, most folks were born and died in the same town or village. However, that kind of world is disappearing, and today's populations are becoming mixtures. The world is smaller these days and is likely to get smaller in the future. Foreign immigrants now abound in most countries. Eventually, the descendants of immigrants will marry the locals, and their offspring will experience gene flow. Compared to the typical local or the typical immigrant, those children are the product of an unusual gene mixture and will have abilities different from them. This trend is of ever-increasing gene flow accelerates human evolution.

The Red Queen Effect Is Always There

While it is true that we no longer have to worry about a leopard making a dinner of us, microscopic invaders continue to be an ongoing threat. Malaria still kills over one million people every year, mainly young children and mainly in sub-Saharan Africa. Nearly a half billion people suffer from the parasitic disease each year. Influenza is still a fearsome killer, sometimes striking with unusual effect and killing tens of millions of people. Bacteria and viruses have one big advantage over us; they can evolve much faster than us. There exists a never-ending war between our immune system and the microscopic invaders that want to get in. Of course, this is also true for all animals.

That ongoing competitive process is called the "Red Queen Effect" after a story from the Lewis Carroll-book *Through the Looking Glass*. This book is a sequel to *Alice in Wonderland*. The story goes as follows: Alice is running in pace with the Red Queen, who is a living chess piece. Alice says, "Well, in our country, you'd generally get somewhere else—if you ran very fast for a long time as we've been doing." The Red Queen responds with, "A slow sort of country!" "Now, here, you see, it takes all the running you can do to keep in the same place. If you want to get somewhere else, you must run at least twice as fast as that!" Evolution often works like this. An arms race develops between predator and prey, where any advantage gained by one side forces the other side to develop a counter measure or go extinct.

Consider the Red Queen Effect operating between humans and microbes. We deluded ourselves into thinking that we could put an end to disease with medical technology. Diseases like syphilis, gonorrhea, tuberculosis, and others were once thought to be eliminated as threats to society. However, the battle never ends, and bacteria have special advantages. They can acquire immunity to antibiotics from other bacteria species in ways multicelled animals cannot. The antibiotic-resistant bacteria have come back in a form where treatment is more difficult or nonexistent.

The Threat of Fellow Humans

It is a dilemma. One of humanity's greatest strengths is our social relationship with other humans. On the other hand, many of our greatest threats come from fellow humans. I do not need to list the instances of man's inhumanity to man because we are all aware of them. We are all painfully aware of them. Our evolutionary history going back millions of years has been influenced by both the benefits of cooperation and the threats of human aggression. A good case can be made for the fact that genetic changes evolved in us over time to reduce our violent nature and to essentially domesticate us. Yet, even after those modifications, we still are genetically wired for aggressive behavior. The problem now is that it is so much easier to kill others that we must subdue that aggressive tendency that drives us in the moment.

There are a multitude of theories on what evolutionary forces caused the Homo lineage brain to triple in size over the original ape-size brain. One of them is that human culture demanded higher intelligence in order to survive, mate, and have offspring. Even among chimps in the wild, a complex hierarchy exists, and misreading the politics can end in disaster. I think our bipedal ancestors, the Australopiths, had even more complex societies than the chimps because cooperation and communication was more essential for survival. When our ancestors became hunters with deadly weapons, such as the hand axe, misunderstandings could have more serious consequences.

Now, we know that evolutionary change is always in response to a selective pressure. Therefore, that phenomenal human brain expansion had to be in response to a culling mechanism. Those who had the intelligence to read their fellow human's minds and the wisdom to react appropriately, lived to pass their genes forward. Those who misread the signs, suffered the consequences. This same pressure still exists today. The threat may not be mortal combat, but it could be whether one has a decent income, health coverage, safe neighborhood, and ability to raise a family. The world population of humans has been growing exponentially and is now over seven billion people. Moreover, the complexity of life due to technological advances increases at an astounding rate too. To survive and thrive today, requires greater and greater intelligence.

New Mutations in Humans around the Globe

Mutations are biologically nothing more than an alteration of our DNA. The genetic code could be changed at one nucleotide due to the damaging effects of a cosmic ray or due to a processing error during conception and embryo development. Mutations can be either harmful, neutral, or beneficial. Harmful mutations usually disappear from the species within a few generations, neutral mutations obviously have no effect, and beneficial mutations have the potential to help us. The point is that mutations of all three types occur at a rare but constant rate. Regardless of whether humankind is thwarting the ills of disease using ever-advancing technology, these mutations keep happening as they have always done throughout time.

One beneficial mutation has the potential to change our human capability. Granted that that super mutation has to pass itself down through the generations and get distributed globally, but a new kind of human is potentially possible. The likelihood of a great mutation occurring is a function of how many human mutations are occurring as we speak. Well it turns out that the number of new mutations is directly related to the human population size. Gregory Cochran and Henry Harpending made this point in their book *The 10,000-year Explosion*. They argue persuasively that not only is human evolution still alive and well, but it is larger than ever.

The human population is now over seven billion people, so the number of beneficial mutations must be at record levels. Despite the great successes of modern medicine over disease, the arms race of viruses, bacteria, and parasites against our immune system continue unabated. These invaders mutate much faster than we do, and they continually attempt to get past our immune-system defense. Viruses and drug-resistant diseases can kill millions when they strike. So any mutation, which gives us humans superior immunity, will spread through the population.

There have been mutations in the human brain in recent millennia, which made us more successful in human social interaction, learning rate, and execution. Fitting in and contributing translate into better pay, better lifestyle, better child care resources, and the like. Such superior genes can proliferate. A few centuries ago, many people were independent farmers and were less reliant on others. The industrial revolution changed that paradigm markedly. Today, we are highly dependent on each other and need good social skills more than ever before. Moreover, the pace of life gets faster and faster. Today's humans must be able to handle the stress of a fast-paced life and of a fast-changing world. Genes that help us do that prevail, whereas the genes of burned-out individuals probably will not.

Jumping Genes and Human Evolution

Tina Hesman Saey wrote an article (Sci. News 5/16/2017), where she examined the role of so-called Jumping Genes in human evolution. It turns out that the genomes of our ancient and not-so-ancient ancestors have been intrusively invaded by viruses called transposons. These viruses were able to hijack the machinery of our genomes and use it to embed copies of their DNA into it. One type of transposon could make multiple copies of itself and embed them into various regions of our genomes and that is why so much of our genome seems to be nonfunctional today. It is often referred to as our junk DNA. Saey refers to these relics as "transposon fossils" because our genomes are, in a sense, historical records of our distant past. Now, so many generations have passed since some of these early invasions occurred, that their relics have been torn to shards by the gene-shuffling process, which occurs during procreation. Why is this important? It is important because we are learning that there is more to our evolution than obvious natural selection events. It appears that those ancient shards have been used by our bodies to improve genome business. For example, hundreds of these shards have repurposed as micro-RNA, which perform a myriad of functions, many of which are unknown. Scientists know of at least 409 times this has happened. One particular type of application of the micro-RNA is to help boost or dampen protein production and gene activity in the cells. It is also known that small RNA agents, derived from transposons, have been reassigned to prevent transposons from jumping anymore.

In addition, protein segments from the transposon fossils have found in use as enzymes. For example, DNA-cutting enzymes RAG1 and RAG2 are encoded from relics of a DNA transposon. These new enzymes have also played a role in making immune proteins to find and kill ever-changing infectious invaders.

Another benefit derived from transposon pieces are strings of DNA code called "transcription factor-binding sites." These entities have been very significant in mammalian evolution because they can evolve into complex gene-regulating switches. For example, a retrovirus called MER41 has been identified in a host of different mammals. We humans have hundreds of copies of it in our genomes. The benefit of it to us today is an enhanced ability to detect and fight infections. In another example, the transposon HERVK has become vital to protecting human embryos against viral infections.

Researchers are also learning how transposons have played an important role in making humans different from our ape relatives. For example, they discovered a retrotransposon dubbed "Coordinator," which is important in turning on the "Human-forming" versions of genes. These discoveries are helping to explain why we humans are so very different from our living ape relatives even though our genes are 98 percent identical. The difference is in the regulatory genes that control banks of other genes and in the switches, that turn genes on and off. It is a level of complexity, which is better understood in light of these discoveries.

In terms of future human evolution, there are processes available in these transposon fossils, which continue to operate today. Humans are constantly changing in thousands of different ways. Those human variations, which bestow advantages, will increase in the generations that follow. Those that bestow disadvantages will disappear.

Technology and the Future Human

Technology is advancing at an unbelievable pace due to better-educated and more specialized scientists and engineers, a wealth of information at their fingertips, and more powerful tools such as powerful computers, ultrafast DNA sequencers, and others. Take "Crispr" as an example. Crispr is the name of a biological technique where a selected sequence can be snipped out from a gene and replaced by a different sequence quickly and easily. With this capability, countless experiments with modified genes can be evaluated and the sought for result rapidly found. Plants and animals with desired traits can be designed in the labs in record time. We don't have to stop at plants and animals though. Genetically modified humans are just around the corner. Would you like your next child to be a redheaded, blue-eyed girl with a 150 IQ? Coming right up.

Bionic Humans

Bionics Already Has a Toehold

The futuristic concepts of science-fiction writers have a remarkable history of eventually coming true. One of their concepts is nearly certain to become a reality and that is the concept of the "Cyborg." The cyborg is a creature that is part human and part machine. There are already numerous examples of cyborgs in today's human population if that definition is taken literally. Pacemakers are such a machine carried in the human body. There are millions of people alive and active today because a pacemaker regulates their heartbeat. Heart pacemakers are old news, right? Yet, what you might not have known, that there are one-hundred-thousand-plus Parkinson-disease patients with brain pacemakers embedded in them. Those pacemakers alleviate their symptoms and allow them to function more normally. Or think of Vice President Dick Cheney, who survived without a heartbeat for many months. He had no heartbeat because his artificial heart operated with a rotary pump. He eventually replaced the artificial heart with a transplanted human one. Speaking of human heart transplants, they are routine procedures today, but I remember when the first heart transplant procedure was done and how short the life expectancies of those early patients were. Today, all kinds of organs are transplanted routinely. We have come a long way.

The science-fiction cyborg is often depicted as a mechanical robot covered with living tissue. It thinks and talks like a human but may have abilities beyond human limits. The movie *Robocop* depicts a once severely injured policeman with mechanical parts replacing much of his body. The restored cop gains a superior advantage over criminals due to his greater strength, speed, and immortality. Today's humans are far short of that description, but we are becoming more like that cyborg than folks in the 1950s. We are not afraid of altering our bodies. Cosmetic surgery has become routine for those who can afford it. How many upper-middle-class daughters today have shapely breasts, perfect white teeth, and a cute little nose, not due to nature, but due to the skills of a plastic surgeon and an orthodontist? Introducing nonliving replacement parts for our bodies has also become routine. How many seniors do you know with metal knees and metal hip joints? New frontiers in bionic replacement are being actively challenged. Perhaps, the biggest driver in bionics research is the need to repair the disabled military veterans returning home with tragically damaged bodies. The Iraq and Afghanistan wars have produced dismembered young men and women due to the enemy's use of explosive devices. The public demanded that something be done to restore function and dignity to them.

Bionics Marches Forward

The cyborgs of the future are not that distant. The field of neuroengineering is on the threshold of huge breakthroughs. Decades of research have led up to this day. The first prosthetic limbs date back to 1000 BC, but they were made of wood and metal and did not bend or move. It was pure fantasy until a few decades ago to imagine that our brains could actually control a prosthetic. It wasn't until the 1970s that experiments began with animals to get the brain to control a mechanical device. By the 1990s we understood that neuron-firing sequences in the brain accounted for limb movements, and we were able to implant dozens of electrodes in animal brains to record the simultaneous firings. By 2011, we had monkeys use their thoughts to control a software avatar. A year later, humans with implanted brain electrodes were able to manipulate a robot arm with their thoughts. The next step is to allow paralyzed humans to walk again through the motorized movements of an exoskeleton controlled by brain waves, which are translated by portable computer into motor signals. When hundreds of paralyzed people are moving themselves among us again, the age of cyborgs truly has arrived.

The science-fiction cyborg has the ideal prosthetic body parts. They are controlled by the wearer's thoughts, and they provide him with sensory feedback and fine tactile control. Think of the human hand as an example. We can pick up a robin's egg without crushing it, or we can use power to twist off a resistant bottle cap. We are currently a long way from being able to provide that kind of prosthetic, but we may get there in a generation or two. As you read this, bioengineers are working hard to develop prosthetics, which can be controlled by the user's brain and which provide sensory feedback to the user, but it is extremely challenging. One big problem is the interface. Mechanical devices and controls use electricity to drive motors, run computers, and connect sensory signals. Living nerve tissue uses an entirely different method of connecting brain and limb. It relies on chemical impulses. The problem is how to connect the biological and electromechanical signals so that they work together.

You crawl before you walk, and you walk before you run. So it is with bioengineering. If you are unable to make the ideal prosthetics at the present time, what can you do with current technology? The problem of walking is easier to solve than the more complex problem of a working arm and hand, so the walking problem is likely to be solved first. One approach being considered for paraplegics is to fit them with an exoskeleton, which can move their limbs with motors, maintain their balance, and control their gait with computers. Meanwhile, research has been going on for decades on providing the user mental control of the prosthetic device. One method is to reroute auxiliary nerves to control the device and then train the user to acquire control of those muscles. A computer is used between the biological nerves and the prosthetic to translate the signals.

Genetically Engineered Humans

In Vitro Fertilization

Superior intelligence is a highly desired trait in these modern times. It is a commodity sought in the competition between nations, in the competition between corporations, and even parents feel good when their children outperform scholastically. What are we willing to do to have more intelligent children? One route that has been discussed is in vitro fertilization (IVF). For example, if parents had ten embryos to choose from and could select the most intelligent embryo, that baby's IQ could be eleven points higher. If this technique were employed over multiple generations, we would be producing geniuses when rated on today's intelligence scale.

Human Improvement via Gene Therapy

If we are patient enough, the numerous mutations occurring throughout the world will make our great-, great-grandchildren far smarter than we are. The problem is we won't live long enough to witness it. Sorry, but we humans are impatient and don't want to wait for generations to go by to gain an improvement to our species. We want it now. Consider the need to cure or prevent Alzheimer's disease, for example. Losing our ability to think and being a burden on others is not the way we want to spend our golden years, yet the odds of it striking us in our senior years are far uncomfortably high. We need to solve these kinds of problems much faster.

Nature's method of improving a species is to make a beneficial mutation and wait for it to become ever more prevalent in future generations. We may be talking about hundreds of years to spread. One way to shortcut that slow evolutionary process is through gene therapy. Essentially, gene therapy is a technique to introduce desired genes into the cells of a body. We previously mentioned that the human version of the FOXP2 gene for speech was introduced into mice embryo, and those implanted genes changed the speech and comprehension ability of the baby mice. Treatment of humans involves introducing altered genes via a carrier, which imbeds it into the patient's DNA. Of course, the safety requirements for using gene therapy to treat humans are a lot more stringent than these are for test animals. Gene therapy for humans is in the very early stages of becoming a common and accepted practice.

So, what is gene therapy? Yes, it is a technique to introduce desired genes into the cells of a body, but it is more than that. Those implanted genes actually have to work once they are there. In other words, they must actually produce the protein needed for the trait that we seek. And for that to happen, we need what is called a vector. It turns out that nature already has vectors that do just that. Two of those vectors are the retrovirus and the adenovirus. So, what is needed next is to strip out those particular genes already in the virus, which make us sick, and then insert the gene or genes that we wish to implant. Finally, the vector is applied to the subject. This can be done by injection or intravenously into the specific tissue to be treated. Alternatively, the subject's cells can be removed, treated in a lab, and reintroduced to the subject.

If all this sounds simple to you, it's not. Viruses are microscopic in size, generally smaller than bacteria. Moreover, the processing is expensive and requires skilled technologists and special equipment. The medical industry is advancing this research to solve otherwise unsolvable problems. Gene therapy is used to treat serious medical conditions, like cancer and hemophilia at this time. Gene therapy may solve the problem of defective bone-marrow cells, or immunodeficiency. In the latter case, where children are lacking the ability to fight infection and must be isolated from exposure to germs, gene therapy has cured 90 percent of early trial cases. Those children will be able to live normal lives. However, unexpected complications also arise with gene therapy. These problems will eventually be solved. We expect that as gene therapy becomes safer, more reliable, and more effective, the price will come down, and it will be used to solve a greater range of medical problems. And in time, it will be used to make humans smarter, stronger, prettier, or whatever changes we desire. We will also find a way to make those improvements inheritable. And when we reach that stage, we will be reengineering the human race.

Advancements in Human-Thinking Ability

Solving Mental Problems

In a TED Talk Presentation in September 2016, neuroscientist Rebecca Brachmann informed us that we humans are in the middle of an epidemic of psychiatric diseases and mood disorders. She tells us that as many as one in four people suffer from some form of mental disorder, yet the best we can do for these sufferers is suppress the symptoms with antidepressants. In the first historic breakthrough in dealing with this problem, Iproniazid was found to be effective for many people in reducing their depression. However, this drug produced severe side effects such as excessive weight gain, liver toxicity, and suicidal tendencies. It also caused the brain to produce high levels of serotonin. Based upon this revealing information, scientists focused on developing drugs that produce serotonin and were also safer to use. Prozac was the result of all that research.

Brachmann tells us that one particular drug, which she has been testing, actually cures or prevents mood disorders for an extended time. Moreover, the drug may potentially prevent stress-induced mental disorders, such as depression and PTSD, from ever occurring. The drug is Calypsol, which is chemically known as ketamine. The drug has been around for decades in use mainly as an anesthetic but has not previously been evaluated as a cure for mental disorders. It does not work by producing serotonin. Instead, it works on glutamine, another neurotransmitter.

This is how the breakthrough occurred: Brachmann had been testing the effects of different antidepressants in mice studies. When she tested Calypsol, she found startling results. Mice injected with Calypsol and subjected to stress did not cower and hide as mice normally do in these circumstances but roamed around their enclosure as if they had never been stressed at all. Brachmann felt these results were simply too good to be true. So she retested and retested several times more. Then she had other labs duplicate her procedures. The results were always the same. It actually worked! Moreover, the effects of Calypsol were long lasting. This is not true for conventional antidepressants.

So how do these mice studies apply to humans? Brachmann is hopeful that one immediate application might be to protect individuals, who go into stressful situations. This could include combat soldiers, police, firefighters, refugees, prison guards, and first responders. She is even cautiously hopeful that these discoveries may spell the end to stress-induced depression and PTSD for humankind. The bigger lesson for us in this story is that we now live in a world entirely different from anything that existed before. Huge changes to humankind are developing unbeknown to us as we go about our normal lives. We humans are capable of amazing things including transforming ourselves. It is impossible to say what the future might bring.

Long-Living Humans

Although we try not to think of it very often, we all are going to die someday. We all hope that that day is far in the future. The good news is that the human life expectancy keeps getting longer. We have been adding two years to the worldwide life expectancy every decade. However, there is a limit to our biological life expectancy, and it is about eighty-five years. By that age, free radicals have degraded tissue and organs to the point where the repair rate is lagging the degradation rate. Of course, some people live longer than that, and scientists are highly interested in what it is about them that buys them extra time. This is the next frontier. Soon, everything will be done to reduce infant mortality, pestilence, and diseases. Health services will keep us alive until we are old, but then we face that biological limit.

Perhaps, we should look to bats! Bats live a challenging life, needing to eat frequently or starve, and needing the vigor to fly while employing echolocation to navigate a dark world. Despite these challenges, bats live longer than any other mammal accounting for size. What is their secret? Can we learn from it and apply it to extending our lives?

References

Chapter 1

Canup, R. M. "Forming a Moon with an Earth-like Composition via a Giant Impact", *Science* 17 Oct 2012:1226073 DOI: 10.1126/science.1226073

Krauss, Lawrence M. *Why There Is Something Rather Than Nothing.*

Chapter 2

Figure 2.1 is a composite of several images. Credit for the individual images is as follows:

The Aysheaia:

By Esv - Eduard Solà Vázquez (Own work) [CC BY 3.0 (http://creativecommons.org/licenses/by/3.0)], via Wikimedia Commons

The Waptia:

By Obsidian Soul (Own work) [CC BY 3.0 (http://creativecommons.org/licenses/by/3.0)], via Wikimedia Commons

The Opabinia:

By Nobu Tamura (http://spinops.blogspot.com) (Own work) [GFDL (http://www.gnu.org/copyleft/fdl.html) or CC BY 3.0 (http://creativecommons.org/licenses/by/3.0)], via Wikimedia Commons

The Wiwaxia:

By Matteo De Stefano/MUSEThis file was uploaded by MUSE - Science Museum of Trento in cooperation with Wikimedia Italia., CC BY-SA 3.0, https://commons.wikimedia.org/w/index.php?curid=48252133

The Anomalocaris: By Yinan Chen (www.goodfreephotos.com (gallery, image)) [Public Domain], via Wikimedia Commons

Carroll, Sean. *Endless Forms, Most Beautiful.*

Gould, Stephan Jay. *Wonderful Life.*

Lane, Nick. *Oxygen, The Molecule that Made the World.*

Ward, Peter D. *Out of Thin Air.*

Ward, Peter, and Joe Kirchvink. *A New History of Life.*

CHAPTER 3

Figure 3.6 is a composite of several images. Credit for the individual images is as follows:
Stegosaurus:
By T. Smit [Public domain], via Wikimedia Commons
Brachiosaurus:
By Nobu Tamura (http://spinops.blogspot.com) (Own work) [GFDL (http://www.gnu.org/copyleft/fdl.html), CC-BY-SA-3.0 (http://creativecommons.org/licenses/by-sa/3.0/) or CC BY 2.5 (http://creativecommons.org/licenses/by/2.5)], via Wikimedia Commons
Allosaurus:
By Rlevente, original author: Nobu Tamura (Own work) [CC BY-SA 4.0 (http://creativecommons.org/licenses/by-sa/4.0)], via Wikimedia Commons

Figure 3.7 is a composite of several images. Credit for the individual images is as follows:
Parasaurolophus:
By Tim Bekaert (Own work) [Public domain], via Wikimedia Commons
Triceratops:
Charles R. Knight [Public domain], via Wikimedia Commons
Tyrannosaurus:
By Marcin Polak from Warszawa / Warsaw, Polska / Poland (Tyranozaur Rex Uploaded by FunkMonk) [CC BY 2.0 (http://creativecommons.org/licenses/by/2.0)], via Wikimedia Commons
"Atlas of Life on Earth" by Dougal Dixon et al

Bakker, Robert T. *The Dinosaur Heresies.*
Carroll, Sean. *Endless Forms, Most Beautiful.*
Dixon, Dougal, et al. *Atlas of Life on Earth.*
Lane, Nick. *Life Ascending: The Ten Great Inventions of Evolution.*
Malam, John, and Steve Parker. *Encyclopedia of Dinosaurs.*
Ward, Peter. *Rivers in Time.*

Ward, Peter. *Under a Green Sky.*
Ward, Peter D. *Out of Thin Air.*
Zimmer, Carl. *At the Water's Edge.*

Chapter 4

Figure 4.1 is a composite of two images. Credit for the individual images is as follows:
Carodnia:
By Apokryltaros at English Wikipedia [GFDL (http://www.gnu.org/copyleft/fdl.html) or CC BY 3.0 (http://creativecommons.org/licenses/by/3.0)], via Wikimedia Commons
Basilosaurus:
By Nobu Tamura (Own work) [GFDL (http://www.gnu.org/copyleft/fdl.html) or CC BY 3.0 (http://creativecommons.org/licenses/by/3.0)], via Wikimedia Commons

Dixon, Dougal, et al. *Atlas of Life on Earth.*
Officer, Charles, and Jake Page. *The Great Dinosaur Extinction Controversy.*
Powell, James Laurence. *Night Comes to the Cretaceous.*
Redmond, Ian. *The Primate Family Tree.*
Turner, Alan. *National Geographic Prehistoric Mammals.*

Chapter 5

Falk, Dean. *The Fossil Chronicles.*
Gibbons, Ann. *The First Human: The Race to Discover Our Earliest Ancestors.*
Johanson, Donald, and Maitland Edey. *Lucy: The Beginnings of Humankind.*
Johanson, Donald, and Blake Edgar. *From Lucy to Language.*
Leakey, Richard, and Roger Lewin. *Origins Reconsidered: In Search of What Makes Us Human.*
Lockwood, Charles. *The Human Story.*

Morrell, Virginia. *Ancestral Passions.*
Tattersall, Ian. *Becoming Human: Evolution and Human Uniqueness.*
Tattersall, Ian. *The Fossil Trail.*
Tattersall, Ian, and Jeffrey Schwartz. *Extinct Humans.*

Chapter 6

Berger, Lee, and John Hawk. *Almost Human.*
Fagan, Brian. *Cro-Magnon.*
Falk, Dean. *Brain Dance.*
Falk, Dean. *The Fossil Chronicles.*
Gibbons, Ann. *The First Human: The Race to Discover Our Earliest Ancestors.*
Johanson, Donald, and Maitland Edey. *Lucy: The Beginnings of Humankind.*
Johanson, Donald, and Blake Edgar. *From Lucy to Language.*
Jordan, Paul. *Neanderthal.*
Leakey, Richard, and Roger Lewin. *Origins Reconsidered: In Search of What Makes Us Human.*
Lockwood, Charles. *The Human Story.*
Morrell, Virginia. *Ancestral Passions.*
Philip Rightmire, C. *The Evolution of Home Erectus.*
Tattersall, Ian. *Becoming Human: Evolution and Human Uniqueness.*
Tattersall, Ian. *The Fossil Trail.*
Tattersall, Ian, and Jeffrey Schwartz. *Extinct Humans.*

Chapter 7

Dawkins, Richard. *The Selfish Gene.*
Harris, Eugene E. *Ancestors in Our Genome.*
Paabo, Svante. *Neanderthal Man—In Search of Lost Genomes.*
Ridley, Matt. *Genome.*
Ridley, Matt. *The Red Queen.*
Sykes, Bryan. *The Seven Daughters of Eve.*
Watson, James D. *DNA: The Secret of Life.*

Watson, James D. *The Double Helix.*
Wells, Spencer. *The Journey of Man.*

Chapter 8

Falk, Dean. *Brain Dance.*
Falk, Dean. *The Fossil Chronicles.*
Klein, Richard with Blake Edgar. *The Dawn of Human Culture.*
Skoyles, John R., and Dorian Sagan. *Up from Dragons.*
Stringer, Chris. *Lone Survivors.*

Chapter 9

Cochran, Gregory, and Henry Harpending. *The 10,000 Year Explosion.*
Diamond, Jared. *Guns, Germs, and Steel.*
Scarre, Chris, ed. *The Seventy Wonders of the Ancient World.*

Chapter 10

Cochran, Gregory, and Henry Harpending. *The 10,000 Year Explosion.*
Skoyles, John R., and Dorian Sagan. *Up from Dragons.*

Made in the USA
Columbia, SC
31 December 2020

30100032R00153